AIRLINE MAPS

A Century of Art and Design

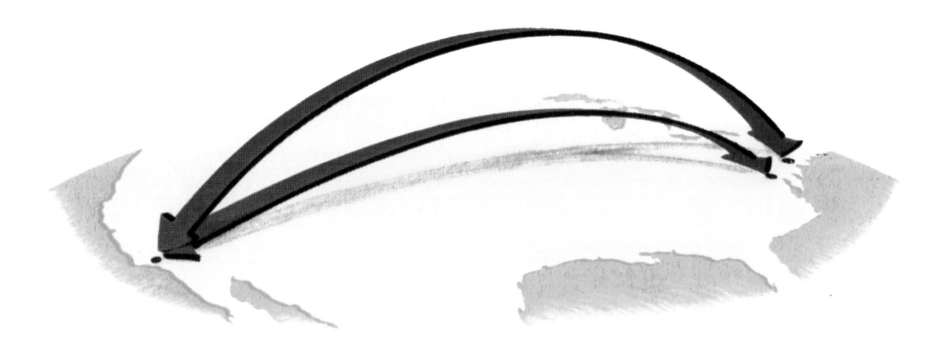

Mark Ovenden and Maxwell Roberts

PARTICULAR BOOKS
an imprint of
PENGUIN BOOKS

Acknowledgments:

Habib Aidroos, Bill Allen, Freja Annamatz, Helen Barnes, Clasonda Armstrong-Grandison, Mike Ashworth, Gklavas Athanasios, Eytan Atias, Cameron Booth, Natalya Borisovskaya, Matthew Brellis, Christian Van Buggenhout, Pat Chessell, Mark Cox, David Crotty, Jim Davies, Martin Dodge, Paula Donaghy, Bradford Drazen, Parker Field, Gayna Fitzgerald, Marie Force, Stephen Gay, Jonathan Gesmundo, Debashis Golder, Sietske De Groot, Phil Harper, Tom Heitzman, Consuelo Arias Hernández, Kate Igoe, Inka Ikonen, Neda Jaafari, Dick Janson, Rohandra John, Jon Kallwejt, Christina Kobel, Daniel Kusrow, Bjorn Larsson, Alexandra Martins, Chris McCloud, Holly Mitchell, Lára Margrét Möller, Pramila Banymandhub Nowbuth, Gerard O'Shea, Denis Parenteau, Gordana Pavlovic, Laurence Penney, Olivier Gilles Rebours, George Ritchie, Andrea Rodriguez, Linnette Romero, Carolyn Russo, Christian Rysler, Chris Saynor, Hannah Searle, Rob Shepherd, Valentina Taddeo, Nuttavika Tamthai, Michael Walton, Barry Weekley, Karl Wuethrich.

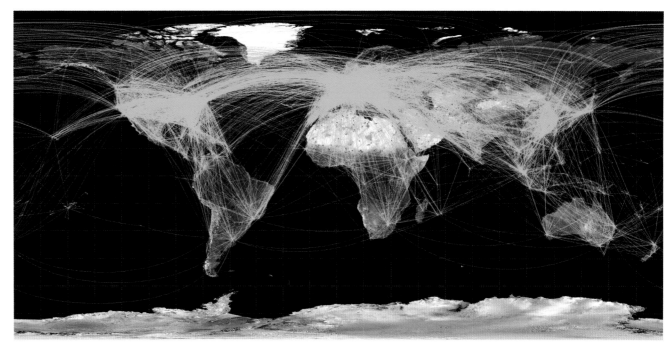

ABOVE: *These aircraft tracks from December 2016, recorded by openflight.org in a twenty-four hour period, are so numerous over the United States, Europe, and parts of China that they appear to obliterate the land beneath.*

PARTICULAR BOOKS

UK | USA | Canada | Ireland | Australia
India | New Zealand | South Africa

Particular Books is part of the Penguin Random House group of companies whose addresses can be found at global.penguinrandomhouse.com.

First published in the United States of America by Penguin Books 2019
First published in Great Britain by Particular Books 2019
001

Set in Frutiger 45, 64, and 75
Designed by Mark Ovenden
Printed and bound in Italy by Printer Trento S.r.l.

A CIP catalogue record for this book is available from the British Library

ISBN: 978–0–241–43412–3

www.greenpenguin.co.uk

Image Credits:

The majority of maps have been sourced from the collections of the authors, British Airways Speedbird Centre (with thanks to Jim Davies), and www.timetableimages.com (with thanks to Bjorn Larsson). Images have been reproduced with the kind permission of the airlines who produced them (or their designers). Credits for specific images are listed below. Extensive effort was made to contact all other copyright holders.
The authors are also indebted to the following for their contributions to this book:

Multiple images: Aeroflot, pp. 71, 91, 113, 116, 117. Air Canada, pp. 75, 83, 103, 110, 114, 121, 125, 141. Air France Museum, pp. 8, 10, 21, 23–27, 35, 38, 40, 41, 53, 62, 63, 88, 89, 94, 95, 97, 98. Air India, pp. 76, 77, 85, 100. airline-memorabilia.blogspot.com, pp. 105, 112. Alitalia, pp. 9, 19, 21. American Airlines, pp. 4, 43, 50, 57, 66, 111. Avianca, pp. 59, 123. British Airways Speedbird Centre, pp. 5, 10, 11, 21, 23, 34, 36, 46, 47, 49, 54, 55, 61, 64, 65, 72, 73, 80, 82, 98, 100, 101, 104, 112, 130. Czech State Airlines, pp. 20, 22, 45, 94, 127. The David Rumsey Map Collection at Stanford University, pp. 4, 13, 16–17, 24–25, 31, 42, 43, 47–49, 51, 52, 53, 56, 57, 72, 73, 81, 84, 88, 89, 95, 99, 131. Delta Flight Museum, pp. 13, 15, 17, 30, 31, 33, 36, 39, 46, 51, 64, 96, 100, 101, 106, 107, 122, 124, EasyJet, pp. 118, 136. Egyptian Airways: pp. 18, 20, 60, 92, 139. Hawaiian Airlines, pp. 84, 108. Iberia, pp. 90, 97, 105, 127, 133. Intourist, pp. 30, 128. Iraqi Airways, pp. 41, 69. JAL, pp. 68, 106, 125. KLM, pp. 5, 12, 14, 16, 19, 23, 27, 58, 87, 94, 98, 115, 126–128, 131, 138. Lufthansa, pp. 12, 44, 101, 112. National Air and Space Museum Smithsonian Institution, pp. 9, 18, 32, 37. Pan Am (courtesy of Delta Flight Museum), pp. 17, 30, 31, 36, 39, 46, 100, 101, 122. Philippine Airlines, pp. 106, 107. Qantas, pp. 47, 60, 135. Kevin Reinhardt, pp. 2, 134, 135. Ryanair, pp. 119, 120, 136, 137. SABENA (courtesy of CVB Bankruptcy), pp. 12, 22, 37, 41, 107. SAS, pp. 78, 81, 109. Saudia, pp. 123, 124. Swissair / Swiss Air Transport Company / SAirGroup, pp. 129, 130. TAP, pp. 59, 120, 129. timetableimages.com, pp. 13, 20, 23, 37, 70, 71, 76, 77, 91, 92, 94, 102–103, 114, 123. TWA Museum, pp. 47–49, 56, 65, 99. United Airlines, pp. 33, 43, 75, 114. **Single images:** p. 3: SAS International. p. 4: Ovenden; Keith Doyle. Screen; flightpath3d.com. p. 5: Arshile Gorky Foundation, photo; Guy Slatcher. p. 7: Allen Airways Flying Museum. p. 15: Kusrow Collection. p. 16: Barry Weekley. p. 20: Library of Congress. p. 64: Ashworth Collection. p. 67: United Aircraft Corporation. p. 68: CAT; Central Intelligence Agency. Thai Airways. p. 69: Athanasios Collection. p. 74: Lockheed Martin Corporation. p. 75: British Columbia Library. p. 79: CAAC. p. 87: El Al Israel Airlines Ltd. p. 92: Ethiopian Airlines, Sudan Airways, Tunis Air. p. 98: Aer Lingus. p. 102: Boeing Corporation. p. 105: Royal Air Maroc. p. 111: JAT, Interflug. p. 112: Bulgarian Airlines, Garuda Indonesian Airways. p. 114: LACSA. p. 115: IranAir. p. 116: TAROM. p. 120: Bahamasair. p. 121: ValuJet; Southwest Airlines. p. 125: Caribbean Airlines. p. 126: LOT. p. 127: Schiphol Airport. p. 129: Swiss International Airlines. p. 134: Boom Supersonic. p. 136: FinnAir. p. 137: Virgin Group, Seattle-Tacoma International Airport / Port of Seattle. p. 138: Kenya Airways, Afriqiyah Airways. p. 139: Emirates. p. 140: Kallwejt Studio. p. 141: Air Mauritius. p. 142: AirEuropa. p. 143: Icelandair.

Departures

Contents

A new branch of cartography

Mark Ovenden

Author, broadcaster, consultant, lecturer

A new branch of cartography took off with the dawn of civil aviation a century ago. Airline maps were completely different from what came before, such as navigational maps used by pilots or even railroad route maps for train riders. Airline passengers did not need to know many specific details about their journey. All they were involved with were the airlines' departure and arrival cities. To those with a fear of flying, perhaps the lack of detail was a clever psychological trick to entice them onto the fastest-moving machine ever invented.

Such a laissez-faire attitude to geographical accuracy gave rise to great experimentation, as the pages in this book testify. The very concept of whizzing rapidly from one side of the world to the other provided artists and marketers with inspiration. During the infancy of aviation, service maps were able to display in an instant both the proximity of the "colonies" and the viability of the new-fangled airlines over any surface-based transportation (especially the great oceangoing liners). Maps played a part in glamorizing flight and making it feel practical, and as networks matured, they were a valuable weapon in beating the competition.

Yet airlines did provide window-seat travelers with aids to assess what they were passing over, as shown in the image to the left, and in the twenty-first century the live map on a seat-back screen or even on a handheld device, shown below, is evolving rapidly. So much so that, in recent years, the tendency of airline websites has been to offer a "point-to-point" drop-down menu with the route laid over something like Google Earth, depriving users of the ability to browse the airlines' wider offering of destinations on a well-designed map. Luckily for the collectors and students of modern industrial design, corporate identity, and cartography, printed paper maps live on. Keep hold of the next one: it could be collectible.

ABOVE: *Passengers consult a 1939 American Airlines brochure.*
LEFT: *Western Air Lines' 1953 guide "Follow Your Flight."*
BELOW: *New scalable seat-back screens by flightpath3d.com have live route maps with night and flight deck views.*

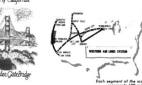

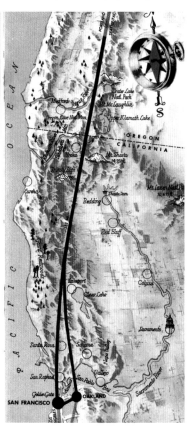

Why airline maps are special

Maxwell J. Roberts

Lecturer; owner of tubemapcentral.com

When we were planning this book, we knew that there would be no shortage of material to include. Our research included surveys of museums, archives, collectors fairs, and enthusiast websites. The more we searched, the more we unearthed. We were astonished by the sheer quantity of airline service maps created worldwide, for publicity posters, in-flight magazines, timetables, and newspaper/magazine advertisements.

Are there more designs of airline maps than railway maps? It is hard to know for sure. Every London Underground station has a network map on every platform, consulted daily by bemused travelers, and millions of copies are printed annually to be taken away and studied at leisure. How many

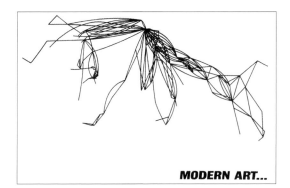

MODERN ART...

people have ever used an airline map to plan a journey? So much cartography going to waste, so why do airlines bother with maps at all?

Because maps don't just show people the way. In the early days of flight, they showed that the impossible was now achievable, and that trips across the world that previously took weeks could now be covered in days. As technology became more reliable, and airline empires expanded, so the maps staked claims on territory, promising regular flights to all manner of exotic destinations. Fly with us—we're bigger, faster (and safer?) than anyone else; travel on a business trip to save time and money, or take a vacation somewhere you never imagined possible. With these messages to convey, it is not surprising that maps became associated with exciting imagery, and as technology, business fortunes, design techniques, and fashions changed, so did the appearance of airline maps. This book tells, in its own way, the story of twentieth-century graphic design.

Airline maps are also special because the cartographer, free of the practical constraints of roads and rails, really does have a blank slate. Only the destinations and intermediate calling points matter, not the actual routes. Although flights are restricted to air corridors and airways, no passenger is going to quibble with the trajectories shown on a map. These can be as short and direct as the designer desires, any shape at all, but with one challenge: distances are often so great that the curvature of the Earth matters, and conveying speed and directness sometimes requires clever distortion of space or choice of projection.

ABOVE: *Cover of a 1928 KLM timetable booklet.*
BOTTOM LEFT: *By 1936, airline maps were understood enough to be turned into abstract art by Arshile Gorky.*
BOTTOM RIGHT: *BOAC routes turned into "modern art"—part of a stylish ad of its services from a 1962 magazine.*

The focus of this book is the airline service maps that publicize operators' empires. Airline route maps, showing flight paths drawn over intricately detailed terrain, play a different role and are less stylistically diverse. They are also under threat from digital cartography; why print a map when a screen can show real-time location? But airline service maps are safe for now. Like all other forms of passenger transport, airlines can only survive if they can persuade the public to travel, and the creativity of designers of airline maps and publicity, in rising to this challenge, flies off every page of this book.

Terminal 1: Aviation takes off

The dawn of passenger air flight came soon after the conclusion of the First World War, with the first scheduled regular international routes established in 1919. Despite the dangers of fragile wood-and-fabric aircraft venturing out over impossible distances, services and maps developed quickly. Two contrasting cartographic themes were established early on, as beautiful, detailed pictorial maps rubbed shoulders with simplified diagrams, embellished with eye-catching graphics.

BACKGROUND: *A rudimentary diagram of all known air routes across the United States, published in 1922 by Bell Telephone.*
LEFT: *Airco de Havilland DH 16 at the 1919 Paris Aero Salon. The model is wearing an early aviator's suit by leading British fashion designer Lucy Duff-Gordon.*
OPPOSITE: *This 1919 poster from Lignes Aériennes Latécoère shows its routes to France, Spain, Morocco, and Algeria. Although the service was provided to deliver mail, passengers were also accepted—hence the advertisement—which makes this the world's first route map for a scheduled international air service. Some of these were printed lithographically on metal, but the ones on paper, like this one printed by Bernard Sirven in Toulouse, France, had a panel below for local details to be overprinted or pasted on.*

LIGNES AÉRIENNES
LATÉCOÈRE

France
Espagne
Maroc

TOULOUSE

Perpignan

Barcelone

Valence

Alicante

Malaga

Gibraltar

Tanger

Ceuta

Larache

Oran

Alger

Rabat

Casablanca

ABOVE: *French aviator Louise Faure-Favier set a speed record between Paris and Dakar in 1919 and was one of the first airline passengers on a commercial flight between the French and British capitals. She wrote this guide to crossing the English Channel in 1921 and went on to make the first live radio broadcast from an aircraft, flying above Paris.*

RIGHT: *A 1921 poster for Transports Aeriens Guyanais, which had only been in existence for three years. Five seaplanes operated two regular routes, both from Saint Laurent du Maroni—one to Cayenne and the other to Inini. Both were a distance of around 155 miles and covered in just two hours, compared with days via boat or over land.*

The rapid technological development of powered aircraft flight is remarkable. Just eleven years elapsed between the Wright brothers' successful experiments in 1903 and the first regular scheduled service in 1914, between two cities in Florida, St. Petersburg and Tampa. Railroad passengers had to wait twenty-six years for equivalent progress with steam

ABOVE: *A relatively early depiction of a streamlined seaplane serving Italy's islands and the Adriatic coast. The poster is thought to date from around 1927.*

RIGHT: *Although Transadriatica was founded in 1925, its first route (Rome–Venice) didn't operate until early 1926. Vienna, advertised here, was next, and within a couple of years Transadriatica went to Brindisi and Munich. Società Aerea Mediterranea took the company over in 1931, which in turn merged with others a few years later to form Ala Littoria.*

locomotives. The First World War, from 1914 to 1918, encouraged further massive improvements, as airplanes provided a new way to attack the enemy and a means of defense. At the end of the war, large numbers of underutilized aircraft and trained pilots created the perfect environment for civil aviation to flourish. The United States Postal Service commenced airmail operations in May 1918, with aircraft capable of 80 mph and, crucially, the ability to fly over

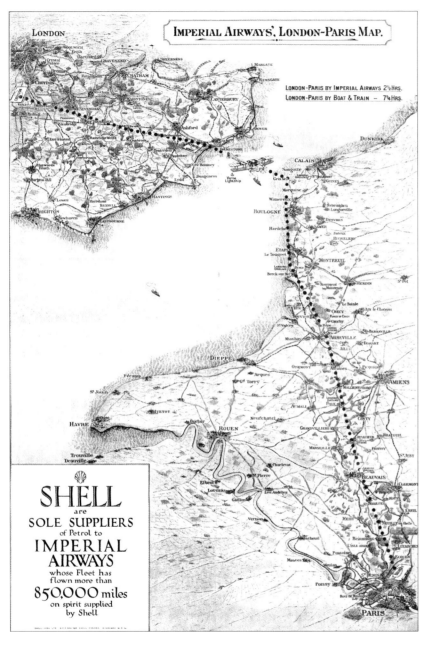

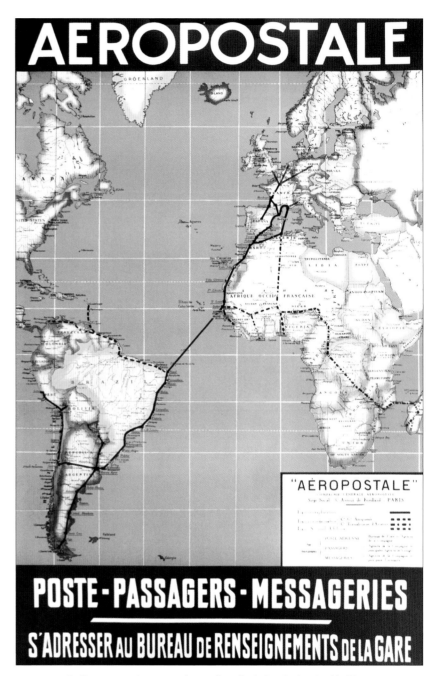

mountains. French airmail service (Société des lignes Latécoère, a precursor to today's Air France) commenced in December 1918 between Toulouse and Barcelona, but passengers had to wait a little bit longer for regularly scheduled international flights. In August 1919, Aircraft Transport and Travel (one of many components that merged together over time to form British Airways) ✈ **14**

ABOVE LEFT: *Shell was an early monopoly supplier of aviation fuel and added its name to many maps, like this 1926 one of Imperial Airways' London-to-Paris route.*
ABOVE RIGHT: *Most early mail services carried passengers too. Aéropostale had been in operation for nine years when this 1927 poster was produced, showing its colonial links.*

ABOVE: *Established with UK government aid, Imperial Airways (an amalgam of four small preceding airlines) was an attempt to build better links to Britain's far-flung Imperial possessions and compete with stronger European companies. This extremely rare surviving example of a designer's artwork, from 1927, shows a proposed Imperial Airways world route map. The finished printed work has not survived (if it was ever produced), but the style resembles a timetable cover that the airline did issue in 1931 (p. 20).*

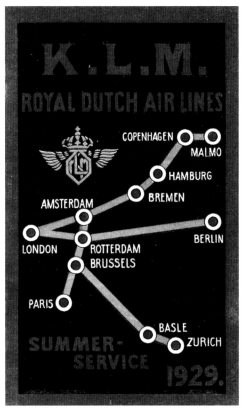

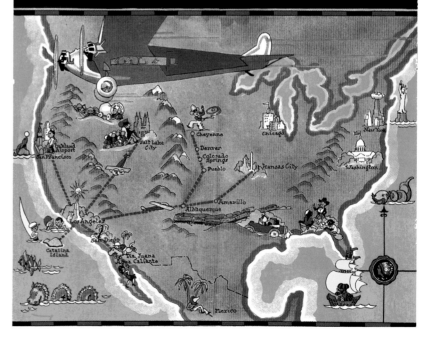

OPPOSITE TOP LEFT: *Just a year after its formation in 1923, Société Anonyme Belge d'Exploitation de la Navigation Aérienne was better known by its acronym, SABENA. This charmingly illustrated postcard map makes no apologies for featuring an entirely female crew and passengers: women having played a major part in early aviation.*

OPPOSITE TOP RIGHT AND BOTTOM RIGHT: *The first flight by KLM was in May 1920 (from London's Croydon Airport to Amsterdam in a DH 16—photo p. 6). Its first intercontinental flight (to Jakarta) took place in 1924. KLM quickly became a major European operator, as shown by these two timetables from 1928 and 1929. By 1930, it was carrying over 15,000 passengers annually.*

OPPOSITE BOTTOM LEFT: *Deutsche Luft Hansa A.G. began operations in Berlin during 1926, inheriting over 150 planes from predecessors and quickly establishing a large pan-European network. Its publicity machine was adept at using color, as this 1929 map shows.*

TOP LEFT: *Established in 1925 as a purely mail carrying operation, Western Air Express was contracted by the United States Postal Service. Within just one month the company was offering commercial passenger services—and evidently making a big play to entice them as early as 1926, using this picturesque map of routes from Los Angeles to as far as Kansas City.*

TOP RIGHT: *Ruth T. White was the illustrator of this beautiful California air services map from 1928.*

BOTTOM RIGHT: *White undoubtedly made the 1929 map of Western Air Express routes too, though she is not directly credited. Her style on all three maps on this page personifies a theme that crops up regularly on airline route maps: beautifully and whimsically illustrated vignettes of local landmarks or people. By twenty-first-century standards, some might be deemed inappropriate, but they are of their time and would not have been intended as offensive.*

ABOVE: *This map dates from 1929 and shows roughly the same stops that KLM Royal Dutch Air Lines made on its maiden voyage to Batavia (modern-day Jakarta) in 1924.*

offered to fly passengers, up to two at a time, from London (Hounslow Heath Aerodrome) to Paris (Le Bourget Airport) in converted wartime de Havilland bombers.

In the early days of passenger flight, few people were willing to pay the expensive fares, but governments quickly appreciated the strategic potential of national airlines. The British soon provided subsidies that enabled its pioneering services to continue and grow, and other countries followed suit. All around Europe, the precursors of many of the world's famous airlines were established, including German Lufthansa (Deutsche Luft-Reederei, 1919), Dutch KLM (Koninklijke Luchtvaart Maatschappij, 1920, the oldest surviving name in air flight), and Russian Aeroflot (Dobrolyot, 1923). The 1920s saw many small operations merging to build up to the eventual giants. For example, Western Air Express, whose beautiful pictorial maps feature in this chapter, was incorporated in 1925 on the back of an airmail contract and merged with Transcontinental Air Transport to form Transcontinental & Western Air Transport in 1930 (eventually

ABOVE: *This detail from a 1929 Western Air Express "Air Log" cover resembles the work of Ruth T. White, who created several maps for the airline.*

TOP RIGHT: *Returning to the style of playful cartoon characters, another Western Air Express route map glorifies the joys of Santa Catalina Island off the coast of Los Angeles, a 1929 playground for Hollywood stars and California's elite.*

BOTTOM RIGHT: *This baggage label from the founding year of Canadian Colonial Airlines in 1929 emphasizes its role as a mail carrier, but passengers were much sought after.*

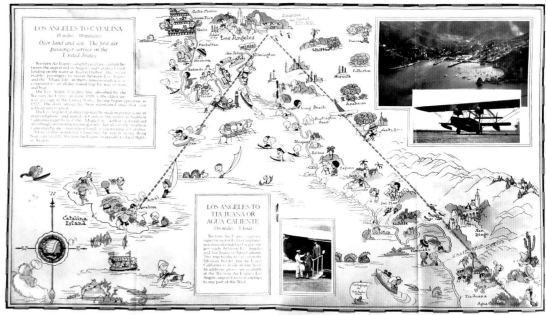

becoming the mighty Trans World Airlines after the Second World War). Other famous names were incorporated directly, later in the decade, with investors keen to cash in on their potential, such as Pan American Airlines (1927) and Braniff Air Lines (1928). The British government had a far-flung empire, from the Americas to Africa to the Antipodes, and it realized that air travel could fulfill a vital role in binding it together. Imperial Airways was formed by decree in 1924, merging the four companies that had been receiving its subsidies.

Given the slow, frail, wood-and-fabric airplanes of the era, ill-equipped for crossing sea and inhospitable terrain, the bravery of pioneering passengers must be marveled at. However, the 1500 deaths on the 1912 *Titanic* disaster would have been world famous. International travel was, simply, dangerous one hundred years ago. Air travel was not noteworthy in this respect, and, with its unrivaled speed, passenger flight was set to take off.

Terminal 2: Empire building

By the 1930s, air services had come of age, with many famous names becoming firmly established. Robust metal aircraft could now convey passengers faster and farther than any other land transport. Commercial passenger flights advertised on travel posters often featured maps as part of the artwork. Air travel was still an expensive luxury, however, and maps combined with sumptuous Art Deco imagery emphasized the glamor, prestige, and excitement of flight. As operators grew, their maps expanded in scope to mark their territories, and global domination became a recurring theme.

BACKGROUND: *KLM had continued its expansion to serve the Dutch colonies through the early 1930s. By 1934, the date of this map, it had introduced the Douglas DC-2 on the Batavia route. That year its aircraft, Uiver, came in second in the MacRobertson Air Race between the UK and Melbourne, Australia.*

LEFT: *Modern painting by Barry Weekley of Imperial Airways aircraft Horatius at London's Croydon Aerodrome (later, Airport) in the early 1930s. When opened in 1920, this was the world's first airport to use air traffic control via its unique control tower (pictured).*

OPPOSITE: *Against a backdrop of German air interests in Columbia, Pan American Airways Incorporated was founded in 1927 as a scheduled passenger (and airmail) service between Havana, Cuba, and Key West in Florida. By taking over various South and Central American companies, Pan Am, as it became known, grew to a dominant position. By the time of this 1938 timetable cover, the company was in its first "Clipper" era and had recently begun transatlantic seaplane operations to the United Kingdom and France, and across the vast expanse of the Pacific to Southeast Asia.*

PAN AMERICAN MAKES THE WORLD *SMALLER*

LEFT: *This striking poster of the 1931 Crociera Aerea Transatlantica (mass flight) between Italy and Brazil was made by Italian artist Umberto di Lazzaro (1898–1968). He produced just over a dozen air-related posters, all of which are eye-catching.*
ABOVE: *Misr Airwork was the only civil aviation company serving Egypt and Palestine when it launched in 1934. Also known as Egyptian Airlines, it soon extended service to Baghdad and Cyprus. The poster is by N. Strekalovsky.*

The 1930s saw rapid consolidation of airlines and (slower) refinement of air travel. At the beginning of the decade, passengers might have "flown" coast to coast by the very first US "transcontinental" service, from New York to Los Angeles. This began in 1929, but the journey involved flight between only Ohio and Oklahoma, and between New Mexico and California, with up to nine passengers squeezed into a tiny Ford Trimotor plane. The remainder of the journey was by train to avoid hazardous night flying.

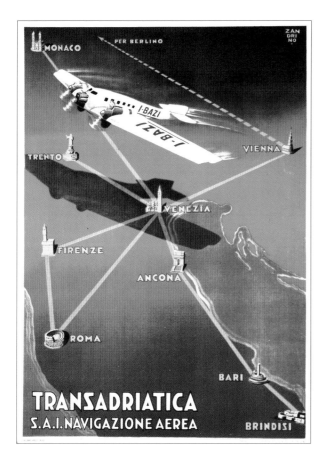

ABOVE: *Painter and illustrator Adelina Zandrino (1893–1994) was better known for her Art Nouveau postcard portraits of women (and later some erotic works), but her 1931 poster for Società Anonima di Navigazione Aerea Transadriatica is one of her finest.*
RIGHT: *As modernism was surfacing, its echoes were occasionally felt in airline maps and posters. The lettering on this 1930 KLM poster evokes the age. The lower panel is set in the Futura typeface (created in 1927).*

The two flight legs shortened the train journey by one day and cost each passenger $338 one way. Developments in safety soon rendered this patchwork journey obsolete. By 1931, the entire air corridor from New York to San Francisco was equipped with radio stations to assist navigation, day or night.

At least the Ford Trimotor took advantage of the latest metal cladding. The frail wood-and-fabric airplanes of the previous decade were quickly superseded by bigger, more substantial aircraft with multiple engines, although the

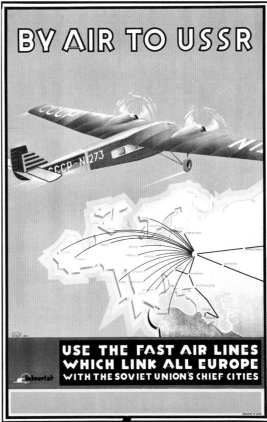

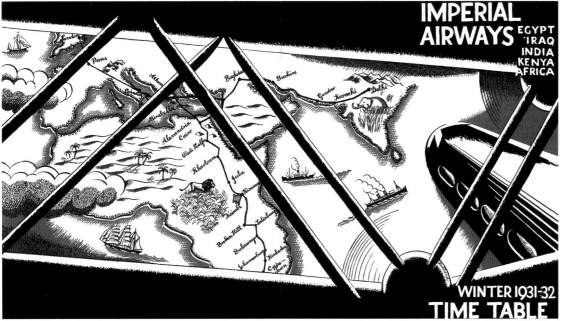

TOP LEFT: *Ceskoslovenské Státní Aerolinie, Czech State Airlines, began in 1923. This 1932 poster uses an early example of "looping," simulating a flight path (including takeoffs and landings), a graphic idea much used in later years.*

TOP CENTER: *The Soviet Union began Intourist in 1929 to give visitors access to travel in the closed country. This 1934 poster is another early example of "looping" routes.*

TOP RIGHT: *Although it closed in 1939—the name was later resurrected—Czech State Airlines produced myriad publicity materials. This 1930 poster was by the artist Vilem Rotter (1903–1978).*

BOTTOM LEFT: *The cover of this 1931 Imperial Airways timetable takes an unusual view of the world, through a plane's wings, although this view would have required the plane to be at lethally high altitude!*

OPPOSITE LEFT: *Although Imperial Airways made use of color, it was not afraid of stylish monochrome, as this 1932 press ad demonstrates perfectly.*

OPPOSITE CENTER: *This 1930 pictorial poster for Aeropostale's La Flèche d'Argent (Silver Arrow) routes is signed by J. des Sachory, but sadly the artist is unknown elsewhere.*

OPPOSITE RIGHT: *Mario Puppo (1905–1970), the prolific Italian poster artist, is the hand behind this 1935 Transadriatica Art Deco advertisement.*

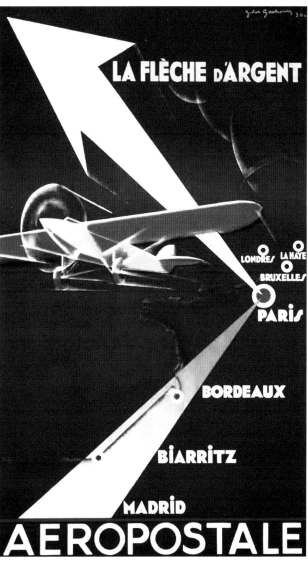

metal cladding inevitably resulted in heavier, more expensive designs that were thirsty for fuel.

Perhaps the most successful airliner of the decade was the Douglas DC-3. Introduced in 1936, this plane could cruise reliably at 170 mph (other civil aircraft of the decade attained 200 mph) and was reputed to be the first passenger aircraft that could be run profitably without a subsidy. It carried a maximum of thirty-two passengers, but many rival models could not even manage half that. With speeds edging upward worldwide, the epic twelve-day, twenty-one-stop journey from Amsterdam to Batavia, mapped in the previous chapter (p. 14), was cut to just five and a half days by 1935, although with the same number of

refueling stops. The intrepid colonial entrepreneur could leave Amsterdam, conduct his business, and be back home before his seagoing rivals had even got halfway there.

Carrying more passengers caused a problem that resulted in a technological dead end. To fly any distance, large heavy aircraft required well-maintained and long runways, facilities that could not be relied upon, especially if refueling was required at a remote outpost en route. Thus were born the massive flying boats of the decade, such as the Boeing 314 aircraft, which could only land and take off in water. Only twelve were manufactured (compared with hundreds of DC-3s), with nine of them purchased just by Pan American for its famous Clipper

 26

TOP LEFT: *Based on an earlier poster by "Olaf" for SABENA, this 1930 version has the London-to-Scandinavia route and the "Baltic Air Express" wording added. The bottom panel was deliberately left blank for overprinting specific tours and prices.*

TOP CENTER: *With a similar trajectory (and extra added stops), this next SABENA, also from 1930, includes a handy compass overlaid on London's Tower Bridge.*

BOTTOM LEFT: *Although some of the typography looks a lot more modern, this is, in fact, a CSA map from 1938.*

BOTTOM RIGHT: *In a more obvious 1930s style, CSA again recruited Vilhem Rotter for this pictorial poster that includes a map inset.*

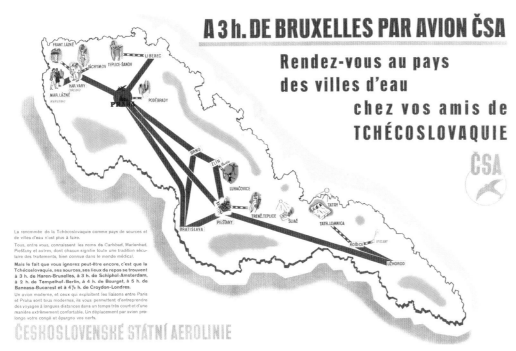

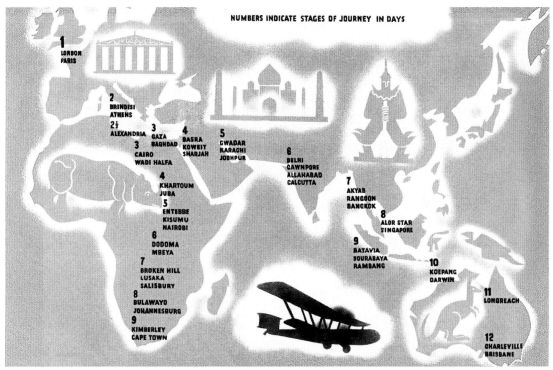

TOP LEFT: *Swedish illustrator Anders Beckman (1907–1967) was a popular choice for travel posters, and, on this 1930 design for KLM, he used the airbrush painting technique for snow-covered peaks.*

TOP CENTER: *This 1933 pictorial poster for KLM is by Japanese designer Munetsugu Satomi (1904–1996), another accomplished airbrush artist.*

TOP RIGHT: *More airbrushing here, this time creating a target effect for a 1933 Flèche d'Orient (Golden Arrow) poster. The route was operated by the Franco-Romanian Compagnie Internationale de Navigation Aérienne (CIDNA).*

BOTTOM LEFT: *This fascinating 1934 airbrushed map from Imperial Airways was designed by poster artist George Chapman (1903–1993). Without any lines of the routes marked, its main job was to show how many days it took to reach the destinations.*

OVERLEAF (pp. 24–25): *Lucien Boucher (1889–1971) produced numerous intricate artworks for Air France and was at his most prolific between 1937 and 1964 (pp. 26, 38, 88). This 1930 work is thought to be his first for the airline—he also produced some illustrated maps for French state railways—and, apart from the occasional winged mermaid here, he mainly focuses on caricatures of buildings and landmarks. In later years his attention turned more to animal representations (p. 26) and people.*

PERCEVAL - 16 rue d'Athènes PARIS

ABOVE: *Expanding Air France operations gave Boucher a chance to widen the canvas in 1934. Aside from the obvious landmarks of China's Great Wall and some gargantuan icebergs, he is clearly enjoying the depiction of animals and indigenous people.*
OPPOSITE TOP LEFT: *Air France was a prolific issuer of publicity, and this 1935 poster not only includes the Hong Kong route but a cutaway of an aircraft (a Potez 600 Sauterelle).*
OPPOSITE TOP CENTER: *Given the date of this Air France poster, it was fairly ahead of the curve in offering "all-in" flight holidays in 1936. The "package deal" became a key 1960s offering.*
OPPOSITE TOP RIGHT: *Even with the latest planes, it was still taking KLM five and a half days to get passengers from Amsterdam to Batavia in 1935.*
OPPOSITE BOTTOM LEFT: *A 1936 timetable cover displays Air France's dedication to allowing experimentation with new and, in this case, unique and abstract concepts for "mapping."*
OPPOSITE BOTTOM RIGHT: *Crossing the Atlantic was hard enough by plane, but traversing the vast open waters of the world's biggest ocean was always going to be tough. By 1937, however, Pan Am was trumpeting its service in a brochure called "Transpacific," in which this marvelous illustration by J. P. Wittlig is center stage.*

services (a flying boat is shown on the timetable cover on p. 17). Up to seventy-four passengers could be carried in the height of luxury across the oceans, with seats converting to bunks at night, and a separate lounge and dining area graced with silver-service banquets. A New York-to-Southampton flight cost $645—not far short of the price of a brand-new Ford V8 automobile—but most flights were transpacific. Owing to their need for suitable water to land in, these behemoths were very inconvenient for rich people visiting fashionable cities like London and Paris, which are far from the sea or extensive lakes.

The more long-term solution was to develop better airport facilities than those offered by the traditional aerodrome, with its grass landing strip and ramshackle buildings. The modern airport, with concrete runways, extensive maintenance facilities, and massive, modern, functional, architect-designed terminal buildings, was born in this era, particularly in the United States, spurred on by World War II.

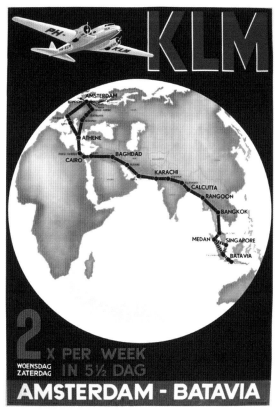

ABOVE: *Attributed to artist Gitta Mallasz (1907–1992), who was born in Laibach (modern Ljubljana), this mid-1930s countrywide plan includes all forms of transport, as the key shows. Four air routes radiate from Budapest, and the exquisite characters, attention to detail, and the joined-up thinking that integrates all methods of travel, merit its inclusion.*

Busier skies also demanded more orderly passage. Croydon Airport, near London, was an early pioneer in 1920, introducing air traffic control, with a control tower for directing all aircraft movements at the airport, and radio beacons for fixing position. The United States was slower to adopt these safety measures; the first airport control tower was not installed until 1930, in Cleveland, Ohio. As another safety innovation, Calgary Airport was the first to be fully lit, also in 1930. Following a number of high-profile accidents, one of which caused the death of a US senator, the first air route traffic control center was opened in Newark, New Jersey, in 1935, followed by Chicago and Cleveland, both in 1936. Their purpose was to organize the movement of aircraft between airports. Thus the foundations were laid for today's airways and air corridors.

 34

AIR LINERS DE LUXE
WITH THE QUIET – COMFORT OF A
LUXURIOUS PULLMAN CAR

SAFETY FIRST
IS DEMANDED BY ALL AIR TRAVELLERS
WE OFFER 100% SECURITY

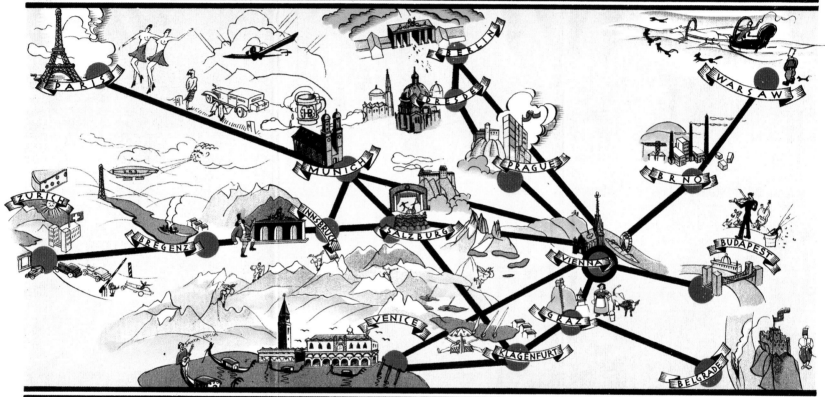

DON'T MISS A TRIP TO VIENNA

THE CROSSROADS OF THE AIR

EAST TO WEST, NORTH TO SOUTH, FROM ONE CORNER OF EUROPE TO ANOTHER
THE NATURAL AERIAL HIGHWAYS MEET AT THE "CROSSROADS OF THE AIR"
— VIENNA IS THE HEART OF EUROPE —
ONCE YOU JOURNEYED DAYS TO REACH US – NOW BY AIR IT TAKES ONLY A FEW HOURS
HE WHO GOES ALOFT TO DAY – WILL GET HIS FRIENDS TO FOLLOW
ALL YOUR FRIENDS KNOW AUSTRIA – DO YOU?
TRAVEL BY AIR — FLY FOR BUSINESS — FLY FOR PLEASURE — YOU SAVE TIME

ABOVE: *With no international frontiers shown, and the ominous date of 1935, readers might be forgiven for wondering if the artist had some portent of what was to come. That aside, the illustrations are whimsical. The authors of this book especially appreciated the red circles, with locations placed in such a way as to be reminiscent of early London Underground signage.*

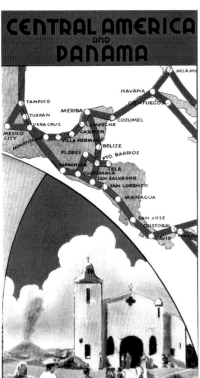

ALL IMAGES ON THIS SPREAD: *Produced in the early 1930s, the era when Pan American was the biggest operator south of the US, this illustrated brochure included sumptuous Art Deco maps on every page. It must have been expensive to produce, reflecting the rich vacation-minded clientele at which it would have been aimed.*

OPPOSITE LEFT: *Pan American Airways had united with the Grace Shipping Company in 1929 to become Pan American-Grace Airways (Panagra). In a bold effort to bring South American capitals closer to New York City, Panagra launched a joint train-to-plane service with the Atlantic Coast Line and Florida East Coast Railway. Passengers were first sent to Miami by rail, and then the Pan Am aircraft took them directly to Havana. The service was expanded to other Latin American countries: by the time of this advertisement for the "Plane Train" in 1930, half a dozen capitals were reached. There was even an aviation passenger station built in Miami.*

RIGHT: *By 1935, when this Pan American pictorial poster was produced, the entire South American continent was served by its "Flying Clipper Ships." Notes by artist Kenneth W. Thompson (1907–1996) include such gems as "Grapes bigger than eggs."*

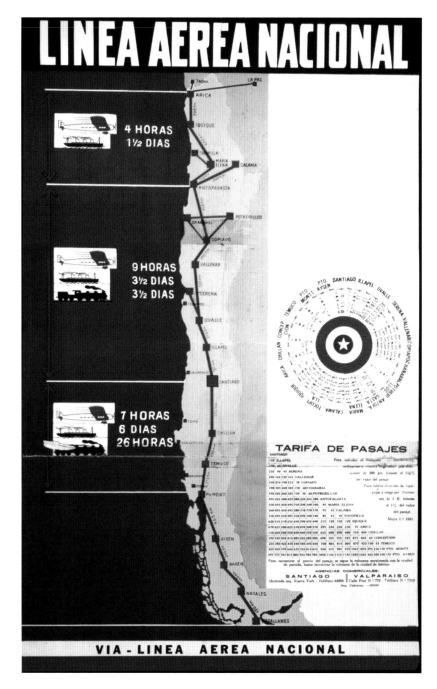

LEFT: *Pan Am was not the only player in South America. This 1937 cartographic poster from Chile's Linea Aerea Nacional includes comparisons between travel time by rail, air, and sea.*
RIGHT: *During the 1930s, Linhas Condor Aereas of Rio de Janeiro was a domestic airline of Brazil, with links to neighboring countries like Bolivia. The elegantly dressed passenger gives a clear statement of the poster's target audience.*

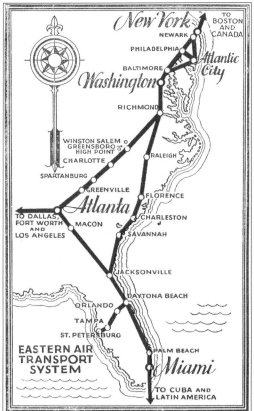

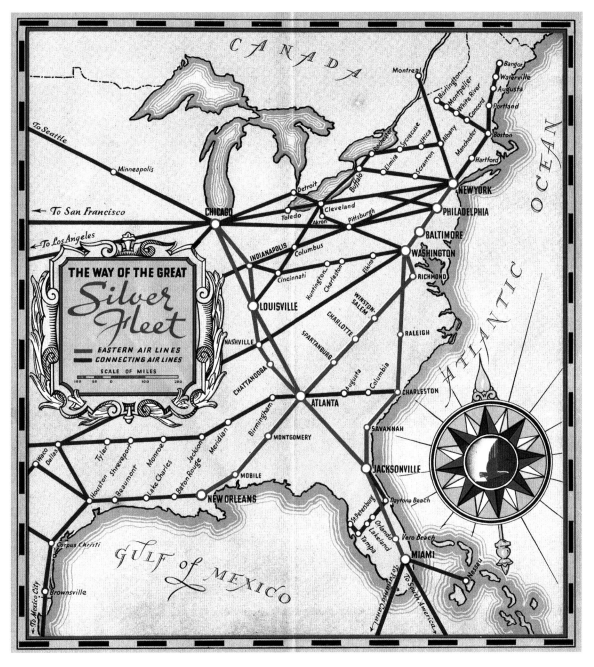

ALL ON THIS PAGE: *All three maps exemplify a simplicity of style, which the authors like to nickname "bus maps."*
TOP LEFT: *United Aircraft Corp. was created by William Boeing in 1928. After several acquisitions, including Varney Air Lines, which enabled it to offer a coast-to-coast service, it became United Air Lines Inc. in 1931. This map dates from 1934.*
BOTTOM LEFT: *Financier Clement Keys merged four lines to create Eastern Air Transport in 1930. This map is from 1931.*
ABOVE: *By 1935, Eastern Air Lines, as it had been renamed, was marketing itself as the "Silver Fleet," with more colorful maps. Bizarrely, this map uses blue for the land and pink for the sea.*

ABOVE: *Imperial Airways had been at the forefront of British industrial design even before Theyre Lee-Elliott had created the Speedbird logo (just visible around the edges) in 1932. This 1935 cartographic poster, with Britain's Empire possessions in red, had the sea (and Speedbirds) in silver, but the map was not easy to read. The pictograms for each city were unique.*

The safety innovations were both timely and essential; the rise in US air travel was explosive during this decade, with around 6,000 passengers carried in 1930, reaching 450,000 in 1934, and exceeding one million by 1938. Even so, air flight was still very much the preserve of the rich, with most people continuing to endure lengthy train rides instead. The seating/bedding and meal comforts offered to the super-rich, however, did not always compensate for the underlying primitive environment. Insulation, heating, and air-conditioning of the fuselage were under the control of the airlines, and they soon claimed to be provided them. Cabin pressurization was not undertaken until after this era, however, starving passengers of oxygen, causing altitude sickness, and preventing the luxury providers from heating water to a high enough temperature for a good cup of tea.

The relatively small aircraft would have been buffeted by turbulence and subject to constant vibration from the engines. To some travelers this would have

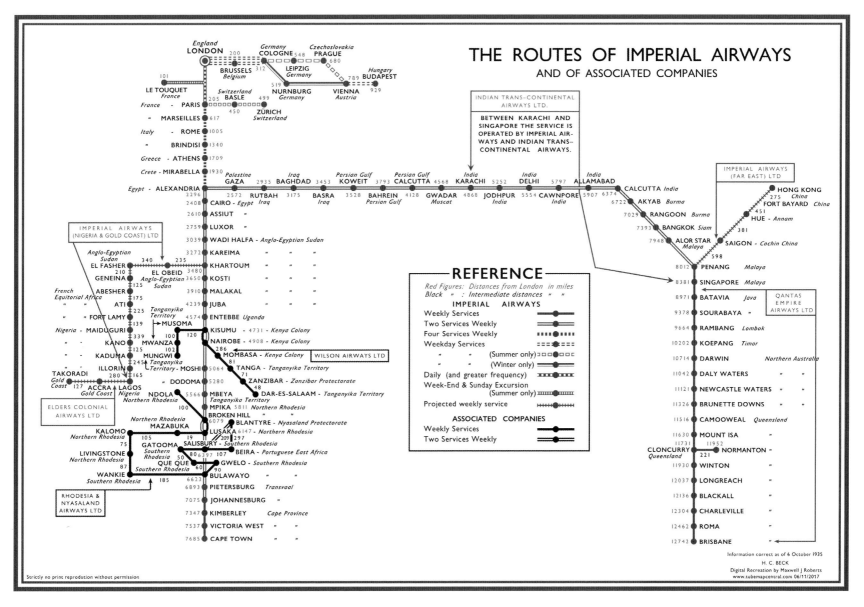

THE ROUTES OF IMPERIAL AIRWAYS
AND OF ASSOCIATED COMPANIES

INDIAN TRANS-CONTINENTAL AIRWAYS LTD.

BETWEEN KARACHI AND SINGAPORE THE SERVICE IS OPERATED BY IMPERIAL AIRWAYS AND INDIAN TRANS-CONTINENTAL AIRWAYS.

IMPERIAL AIRWAYS (FAR EAST) LTD

IMPERIAL AIRWAYS (NIGERIA & GOLD COAST) LTD

WILSON AIRWAYS LTD

ELDERS COLONIAL AIRWAYS LTD

RHODESIA & NYASALAND AIRWAYS LTD

QANTAS EMPIRE AIRWAYS LTD

REFERENCE

Red Figures: Distances from London in miles
Black   "   : Intermediate distances   "   "

IMPERIAL AIRWAYS

Weekly Services
Two Services Weekly
Four Services Weekly
Weekday Services
"   "   (Summer only)
"   "   (Winter only)
Daily (and greater frequency)
Week-End & Sunday Excursion (Summer only)
Projected weekly service

ASSOCIATED COMPANIES

Weekly Services
Two Services Weekly

Information correct as of 6 October 1935
H. C. BECK
Digital Recreation by Maxwell J Roberts
www.tubemapcentral.com 06/11/2017

Strictly no print reproduction without permission

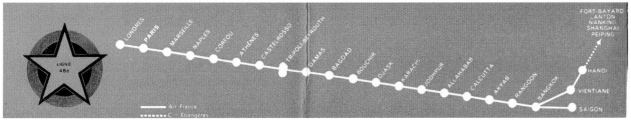

ABOVE: *Imperial turned to the designer of the London Underground diagram, Henry Beck (1902–1974), to simplify its routes. His 1935 work (digitized by book co-author, Max Roberts) is a dramatic contrast to the airline's previous maps.*
LEFT: *Air France used diagrams in its 1936 timetables (p. 27), as this simplification of the Southeast Asia route shows.*

ABOVE: *Elements of Beck's diagram were used on Imperial Airways' iconic 1938 poster.*

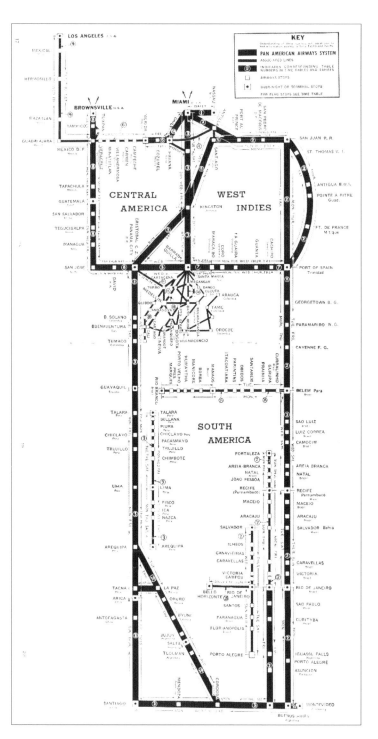

ABOVE: *Pan Am took up the diagram concept with this 1938 timetable insert.*

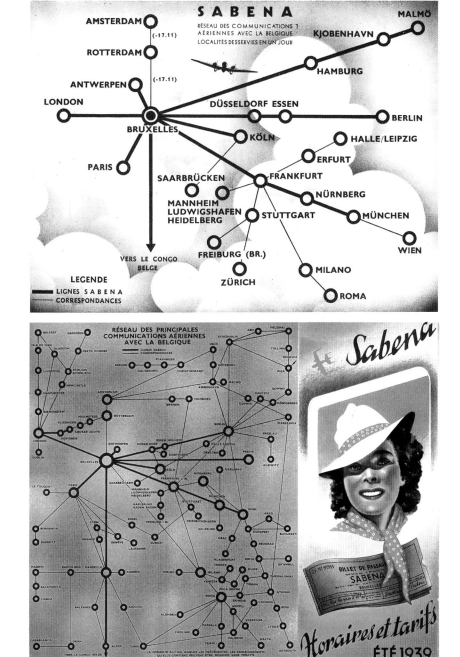

TOP: *SABENA also embraced simplification of its core European routes on this 1937 timetable map, complete with abstract airbrushed spherical clouds.*

BOTTOM: *SABENA's 1939 summer timetable showed a wide range of connections; more details are legible using the diagrammatic concept.*

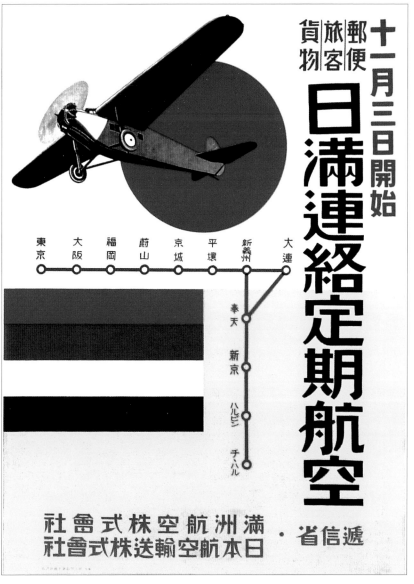

ABOVE: *Did Japanese airlines anticipate the Brits? This poster for a route between Manchuria and Japan suggests maybe they did. It is estimated to date from 1930. The simplicity of the horizontal and vertical lines makes the routes extremely clear.*

been part of the adventure, to others it was intolerable, although sea travel had its own discomforts and dangers. Zeppelins offered comfortable luxury travel and more refinement, but the dramatic Hindenburg disaster of 1937 put an end to their development. In any case, their long-term potential as mass transit carriers of the skies would have been questionable.

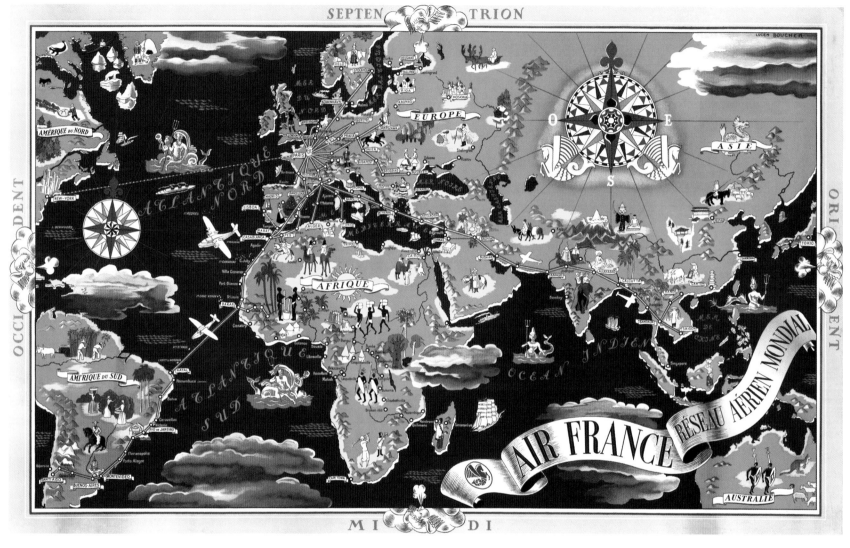

ABOVE: *As the clouds gathered over Europe, artist Lucian Boucher seems to have turned up the dial on color saturation in this 1938 map of Air France's routes. There are now fewer landmarks and many more clichéd caricatures of indigenous people.*

OPPOSITE: *Notice the echoes of Boucher's approach on a 1941 world route map for Pan Am. Artist L. Helguera has excelled for sheer extravagance of the surrounding images, from King Neptune and a bare-chested mermaid, to the aurora borealis and constellations!*

This decade saw the establishment of many famous names in civil aviation. Despite commercial uncertainties (which were countered by government backing in some countries), many companies have since proved to be tenacious and long lasting. From this chapter alone, the names KLM (Netherlands) CSA (Czech Republic), Air France, United Airlines, American Airlines, Lufthansa (Germany), and Qantas (Australia) still grace the sides of aircraft, and SABENA, Pan American, and Trans World Airlines are relatively recent casualties and much lamented.

Imperial Airways, perhaps reflecting an inner belief by the British government that its colonial outposts were on borrowed time, was merged with British Airways (no direct relation to the modern carrier) and renamed British Overseas Aircraft Corporation in 1939. Other famous names originating in this time period include Eastern Airlines, Panair do Brasil, Swissair, Aeroflot (Soviet Union), Northwest Airlines, Northeast Airlines, Delta Airlines, Aer Lingus (Ireland), Avianca (Columbia), and Alaska Airlines.

44

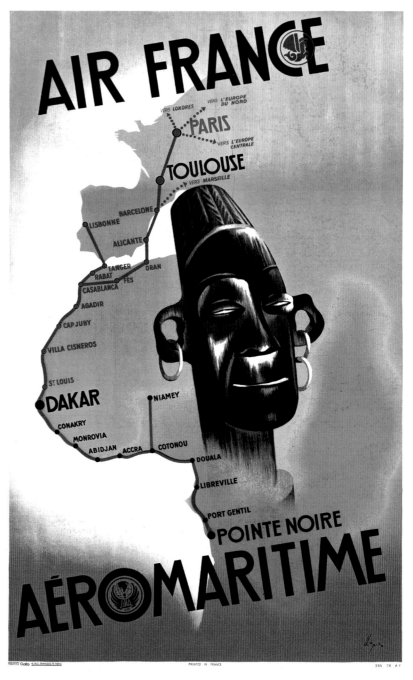

ABOVE: *Another artist hired by Air France was a poster designer who signed as "N. Gerale." It was the nom de plume of airbrush illustrator Gerard Alexandre (1914–1974). This 1939 cartograph, and the next, feature his vibrant signature reds, yellows, and oranges.*

ABOVE: *Also from 1939, and also signed "N. Gerale," this powerful image of carved wooden African art forces the contemporary viewer to address prevailing colonial views of the continent. Gerale's hand-executed Art Deco–esque lettering is also redolent of the period.*

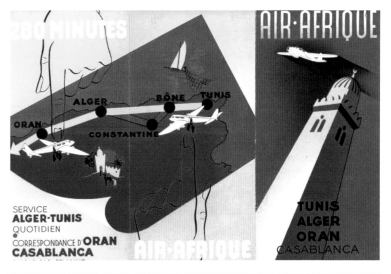

ABOVE: *Belgian colonial interests were well served by SABENA, as this 1939 timetable cover and map show. There is a nice juxtaposition of the landmass shape with its native elephant.*

TOP RIGHT: *The French coastal Maghreb was served by Air Afrique. This 1939 map stresses the speed (just 280 minutes) between Tunis and Oran.*

BOTTOM RIGHT: *Airbrushing was employed by Iraqi Airways on this 1945 poster. There are no lines linking destinations served, a graphic trick that would be used more often in later years.*

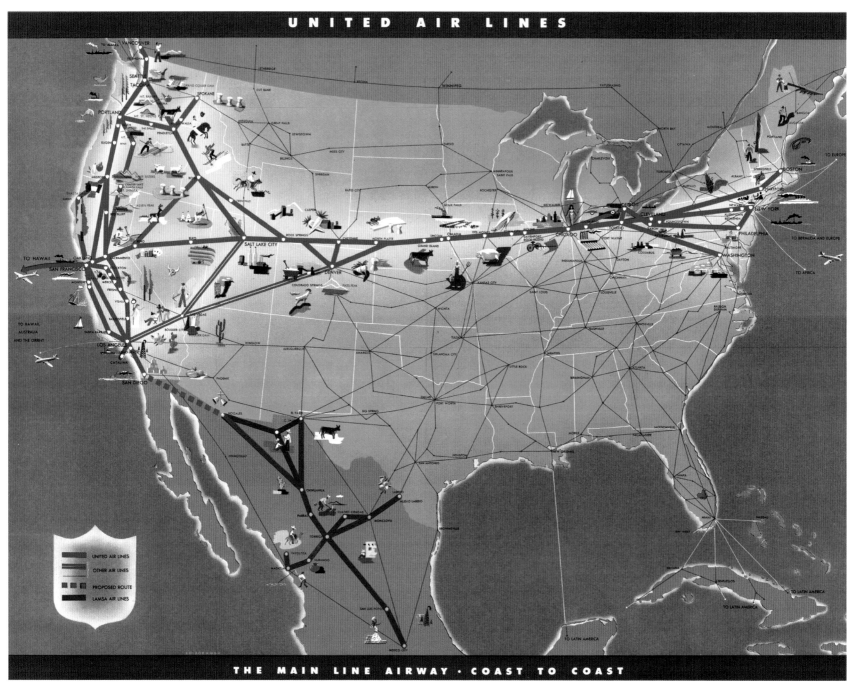

UNITED AIR LINES

THE MAIN LINE AIRWAY · COAST TO COAST

ALL ON THIS SPREAD: *During the war, civil aviation in Europe came to a standstill, but American airlines continued their battle for supremacy in coast-to-coast routes. United Air Lines made great play of its "mid-continent" Atlantic-to-Pacific route (artist Ed Boehmer's colorful 1940 map,* **ABOVE**), *and a more succinct but nonetheless powerful "gay geography"* (**OPPOSITE TOP RIGHT,** *from 1939). American Airlines promoted a more southerly "Flagship" route (1940 ad,* **OPPOSITE LEFT**) *and a nonstop New York-to-Chicago service (1939 map,* **OPPOSITE BOTTOM RIGHT**).

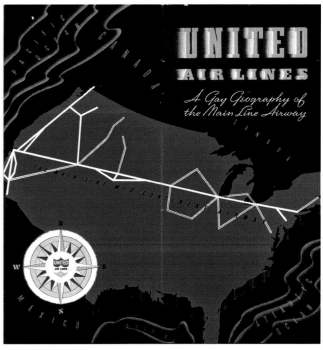

"Just think!..we'll save *two whole days* by Flagship!"

● Add two days to a week to your vacation this summer! Fly American *straight* to your chosen vacation spot in a few comfortable hours, no matter how far away. Even if you're going coast to coast, it's only overnight by luxurious Flagship Skysleeper!

Vacations start at the airport when you Fly American! Superbly comfortable giant Flagships give you the world in miniature to feast your eyes upon as you float smoothly, swiftly along. Rest and relaxation come quickly. Delicious meals aloft sharpen weary appetites. And all too soon— you've *arrived*, ready for play! No vacation ever had a more delightful prelude!

★　★　★

What's your choice—New England, California, dude ranches, evergreen forests, trout streams, the seashore, inland lakes? The historic shrines of Washington, Philadelphia, Boston, Tennessee, Texas, Arizona? All these and more are yours by Flagship. Ask your Travel Agent for complete information and reservations for a vacation by Flagship. Or call the nearest American Airlines office.

AMERICAN AIRLINES *Inc.*

ROUTE OF THE FLAGSHIPS

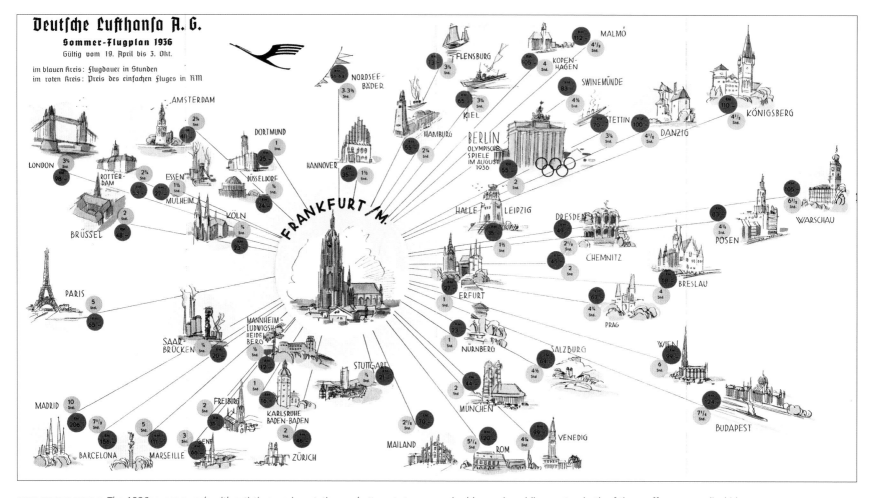

BOTH ON THIS SPREAD: *The 1930s were awash with artistic experimentation and attempts to summarize blossoming airline routes; both of these offer some radical ideas.*
ABOVE: *Showing the flight duration (blue circles) and price of a ticket (red circles) in Reichsmarks (German currency between 1924 and 1948), this 1936 Lufthansa plan includes city landmarks.*
OPPOSITE: *A joint CLS/CSA (Czech airline companies) 1937 diagram also shows flight duration and departure times. A laudable idea, but way too complex to be practical for easy use.*

As befits times in which air travel was a relatively novel indulgence and an exciting pleasure for (at least some) wealthy people, maps of this era are often overtly paired with depictions of the "high life," whether skiing in Scandinavia (p. 23), sun seeking in Central America and the Caribbean (p. 30), or merely relaxing while the porter (presumably) gathers the luggage to take it to the hotel (p. 32). Elsewhere, the perceived glamor of flight is paired with less overt, but no less sumptuous, opulent Art Deco imagery, in its element when conveying the exotic possibilities that would now be only days away rather than weeks (pp. 40–41). On numerous posters and timetables, the airplane is the graphic centerpiece, larger than life, looming over towns and cities, countries, or the entire world, truly shrinking the planet. Even the longest of journeys can be made to look appreciably shorter in this way. With airlines rapidly developing global networks, they sought

to highlight them using grand imagery. Lucien Boucher developed this style in his epic series of dramatic pictorial maps for Air France (pp. 24–26, 38), which inspired impersonations by many other operators with global claims (p. 39).

Given the status of flight in the 1930s, this was generally not the place for the simplicity of modernism. Stark, abstract, undecorated publicity would have seemed wrong. Even so, many of the sumptuous designs used straight lines to depict speed and directness (p. 21), although curved trajectories (p. 20) are scarcely less effective at conveying this. Surprisingly, a few airlines did take simplification to its logical conclusion, depicting their empires as clean, abstract diagrams (pp. 35–37) and evoking a somewhat more mundane subway trip on the London Underground. For modernism in poster imagery, two outstanding examples (CIDNA and KLM, p. 23) demonstrate the use of simple designs to

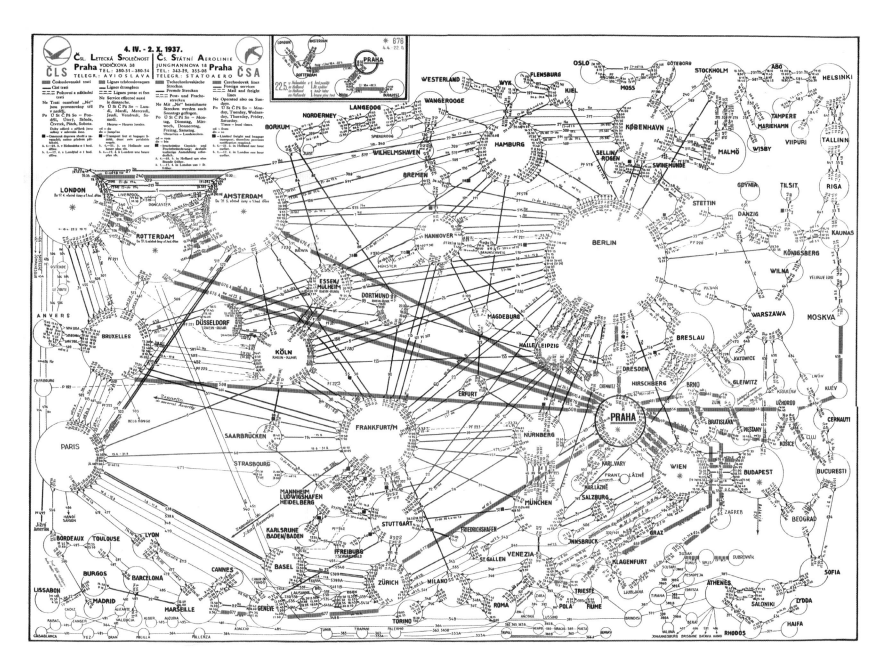

present a clear message, and one of the posters (by KLM) demonstrates nicely the effective use of airbrushing, an artistic technique which came into its element in the 1930s.

Toward the end of this era, with Europe descending into war (and the United States subsequently joining), a distinctly more somber tone is in evidence (pp. 46–47).

The Second World War was set to have massive ramifications for air travel. In the short term, civil airlines in war zones simply ceased to operate, and a six-year hiatus ensued. In the longer term, with the technological and infrastructure developments made during the war, air travel was set to have a massive boost, but minus the exotic glamor and sense of adventure.

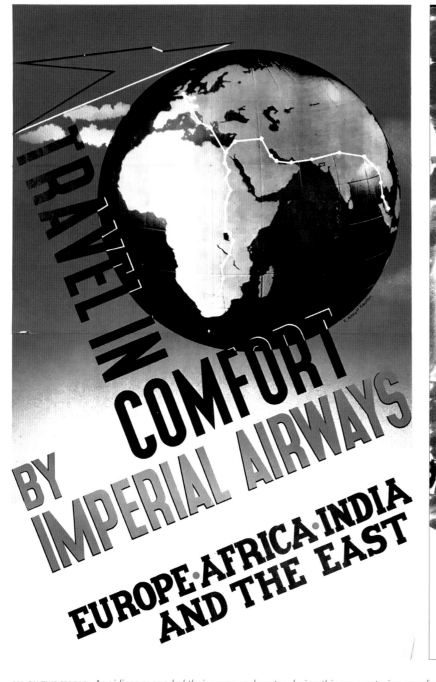

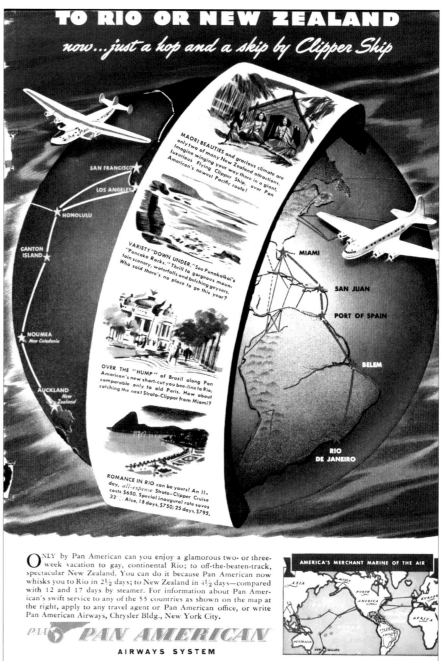

ALL ON THIS SPREAD: *As airlines expanded their range and routes during this era, venturing ever farther from home, the entire globe became their focus of operations. Here it is shown in different ways in posters and magazine advertisements. A 1940 TWA ad was more focused on domestic routes, equating the United States with the globe.*

ROUTINE STUFF

Flying the oceans isn't a new experience to TWA planes and crews. TWA has more than 8,700 overocean flights to its credit, starting in February 1942, with service to Cairo, Egypt. That was the pioneer overocean flight of any domestic airline for the Army Air Transport Command. Later TWA became the first U.S. airline to operate year-round schedules across the North Atlantic. Pictured above is TWA's new trans world system, the foreign portion of which is just as familiar to our crews as are the airways of our domestic routes. Wherever you live along TWA's transcontinental route, you will soon have direct, one-carrier service to key foreign centers halfway around the world to India and Ceylon. If you're going to Europe, Africa or Asia, see TWA or your travel agent now.

NEWFOUNDLAND
IRELAND
FRANCE
SWITZERLAND
ITALY
GREECE
EGYPT
PALESTINE
TRANS-JORDAN
IRAQ
SAUDI ARABIA
YEMEN
OMAN
INDIA
CEYLON
PORTUGAL
SPAIN
ALGERIA
TUNISIA
LIBYA

TWA
TRANS WORLD AIRLINE

IMPERIAL AIRWAYS
AND ASSOCIATED COMPANIES

10½ DAYS

AUSTRALIA

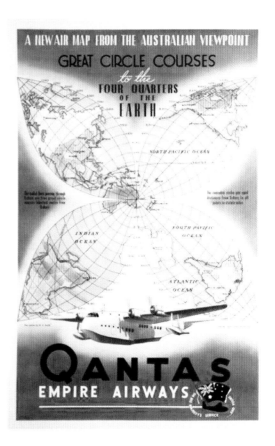

A NEW AIR MAP FROM THE AUSTRALIAN VIEWPOINT
GREAT CIRCLE COURSES
to the
FOUR QUARTERS OF THE EARTH

QANTAS
EMPIRE AIRWAYS

A New World in Travel

OPPOSITE LEFT: *American illustrator Edward McKnight Kauffer (1890–1954) designed 140 posters for London Transport but few for airlines. This very rare one dates from 1935 and displays his love of airbrushing.*

OPPOSITE RIGHT: *Promoting Pan Am's core routes in Central and South America, this 1940 magazine advertisement shows the planet wrapped in a garland of vignettes.*

TOP LEFT: *A 1945 magazine ad from TWA showing how India can be reached via its transatlantic routes.*

BOTTOM LEFT: *A 1940 TWA ad was more focused on the company's domestic routes.*

TOP CENTER: *Artist Albert Brenet (1902–2005) is a widely acknowledged master of gouache; this poster from 1941 for Imperial Airways cleverly implies a globe.*

TOP RIGHT: *Qantas published this unusual Australian-focused butterfly projection in the 1940s.*

BOTTOM RIGHT: *George Cram produced this "Air Age Gingery Projection" in 1943, but it did not catch on.*

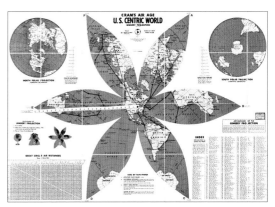

CRAM'S AIR AGE
U.S. CENTRIC WORLD

Terminal 3: Air conquers sea

With the technological developments of the Second World War, passenger aircraft became bigger and could travel farther than ever before. Taming vast distances, transatlantic and transpacific, became a recurring mapping theme, and the rise of the map in magazine and newspaper advertising shows that air travel was no longer the preserve of the wealthy. Pictorial maps depicted the sort of people who might be met in far-off lands, maybe reinforcing unfortunate national stereotypes as a side effect.

ABOVE: *Signed by artist John Fischer, whose beaming caricatures straddle their countries like giants in* Gulliver's Travels, *this 1946 magazine ad by American Airlines promises to "unite" passengers with the "foreign lands" of their ancestors via the "highway of the air."*

OPPOSITE: *Delta Airlines can trace its roots to the world's first aerial crop spraying, in 1924, around the Mississippi Delta. The name Delta Air Service was taken in 1928. By the time of this jolly map in 1946, it had developed a sizable commercial passenger network.*

With the conclusion of the Second World War in 1945, passenger air travel resumed across the ravaged world. It would be several years before the real wartime innovation, jet engines, were successfully incorporated into civil aircraft, but smaller improvements were felt immediately.

Changes in this era were therefore evolutionary rather than revolutionary. Commercial airlines could now purchase large, reworked military transport aircraft that could easily exceed 250 mph. Famous examples include the Lockheed Constellation in 1943, which seated almost seventy passengers.

This became the first civil airliner built in quantity with a fully pressurized cabin, meaning that it could fly over bad weather, increasing passenger comfort and safety. Its major competitor, the Douglas DC-6, introduced in 1946, seated over eighty passengers. Both could make the transatlantic flight nonstop in ideal conditions (refueling at Newfoundland if not), and even made it possible to fly over the North Pole to shorten the distance, with maps (p. 65) emphasizing the new polar route. These advanced aircraft were joined by the Boeing Stratocruiser in 1949, which also seated over eighty passengers, and

56

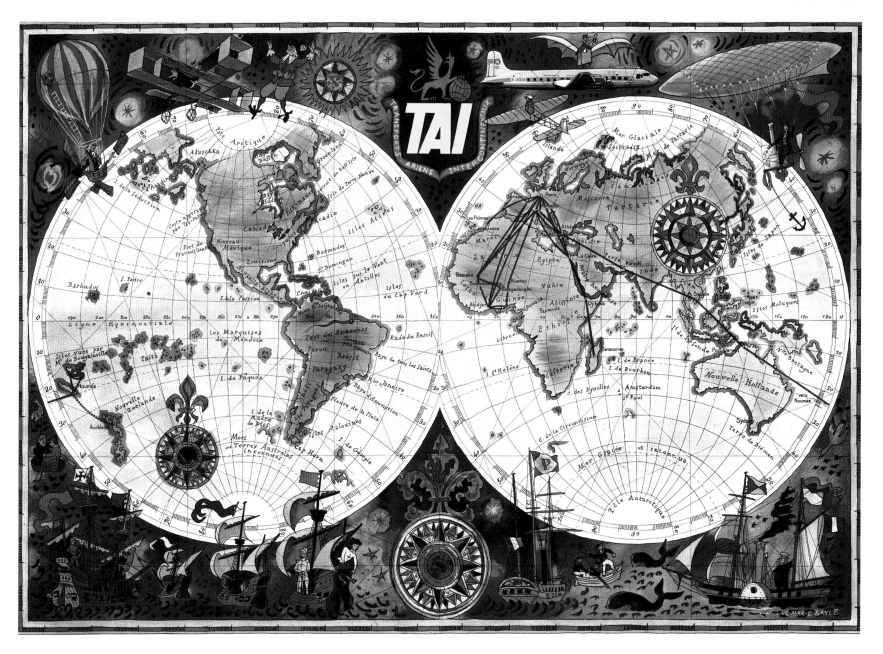

OPPOSITE: *Panair do Brasil, founded in 1929, was a subsidiary of seaplane operators New York, Rio, and Buenos Aires Line, which in turn was part of Pan American. Just after the war, Panair moved quickly to expand from a domestic to an international operator. This could potentially explain the oddity that, while this 1947 map shows half the world and its indigenous peoples, it only promotes a handful of routes between Brazil, Europe, and Africa. The artist, Eymonnet, about whom little is known, clearly borrows from Boucher's ideas, and is exuberant in style with color and borders, but seems slightly less confident with human faces.*

ABOVE: *Transport Aeriens Intercontinentaux began operations in 1946 and, despite commissioning artist Luc Marie Bayle (1914–2000) to depict the entire planet (in Boucher planisphere style), the existing TAI routes at this time (1948) did not serve the Americas. Bayle was a French naval officer with a penchant for illustrated books, but evidently excelled in maritime and aeronautical history, as his delightful vignettes witness. Despite Bayle's optimism of including the Americas, TAI never evolved into a transatlantic operator, preferring to focus on getting to French possessions via eastern routes.*

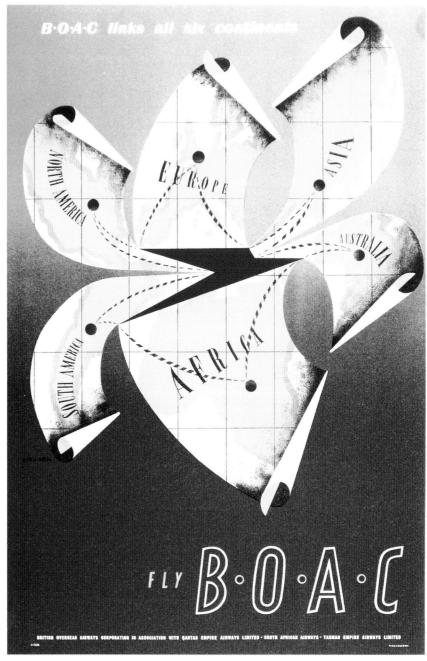

ALL ON THIS SPREAD: *British Overseas Airways Corporation was a state-owned merger of Imperial Airways and the original British Airways companies. Both had high graphic design standards, especially in poster art. In these examples, even the shape of the planet was up for reinterpretation. On the 1947 poster (***ABOVE LEFT***), artist Henri Kay Henrion (1914–1990)—considered the founding father of the modern graphic standards of corporate identity—took liberties with a cube-shaped world. In 1951 the Polish artist Stan Krol (1910–1985) went further and sliced the globe up into a flower with the BOAC Speedbird logo as its stamen (***ABOVE RIGHT***). These exotic ideas helped cement BOAC's reputation at the forefront of UK graphic design.*

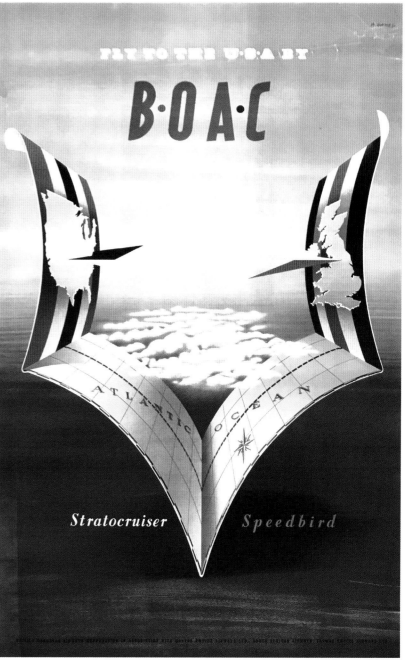

BOTH ABOVE: *Planetary distortion was a recurring theme. Henrion was also inspired by the Speedbird logo, and in his 1947 poster (**ABOVE LEFT**) it was elevated to triumphal levels, flapping around with national flags. The planet is viewed from yet another angle by artist Abram Games (1914–1996), who took distortion even further on his celebrated 1949 poster (**ABOVE RIGHT**), in which the planet is folded away into a chevron, barely showing the Atlantic. What Americans thought of tiny Britain being of equitable size to their vast country is not known!*

T3: Air conquers sea

ABOVE: *On TWA's 1948 map, the directly served cities are shown in script, whereas connections are in sans serif letters. Eagle eyes will spot that Rome is also labeled as Naples, so where the passenger would go for a flight to Cagliari is not clear. The moon and sun, and unserved destinations, like Russia and China, are also depicted by the unnamed illustrator.*

featured optional bunk beds, though installing them would have cut down on passenger capacity.

By the late 1940s, daily flights from London to New York were on offer by Pan Am and TWA. Ready for modern aircraft, London's new Heathrow Airport opened in 1946, and New York's Idlewild Airport, now John F. Kennedy International Airport, commenced commercial flights in 1948. Air travel was

finally in an unassailable position to compete with the ocean liners, both in time and cost. In the mid-1950s, a one-way transatlantic crossing by sea might cost between $300 and $500 for seven days of travel, but TWA was asking just $290 for a one-way flight from New York to London. The erosion in numbers of seafaring passengers was steady but slow, so total defeat did not take place until the jet age.

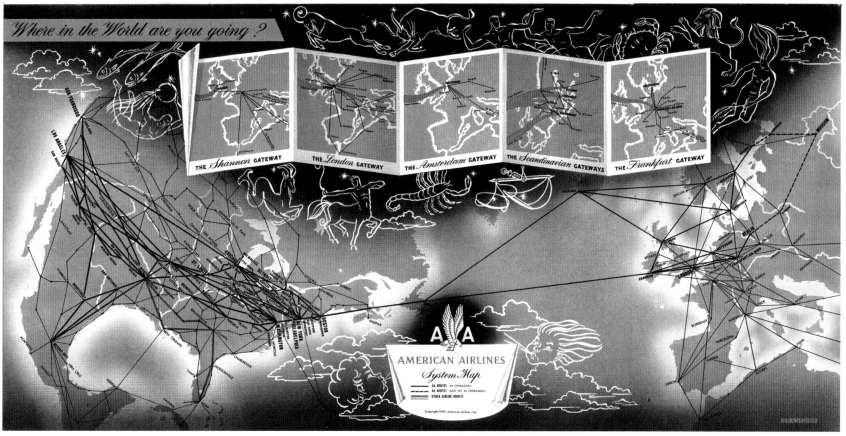

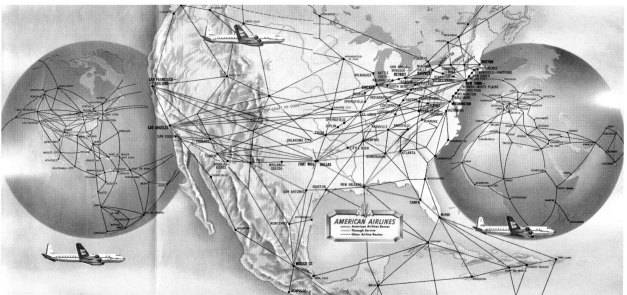

BOTH ON THIS PAGE: *Two maps for American Airlines demonstrate alternate ideas on how to show its global connections. On the 1949 version (ABOVE), North America and Europe are both skewed to one side to allow for a cacophony of constellations and a faux pamphlet of five panels showing its European "gateways" (or hubs, as they are now known). A degree of normality returns in 1954 (LEFT), but here the United States is coddled between two views of a more traditional globe. Both versions display some adorable juxtaposition, but ultimately they also clamor for a more comprehensible way of representing what by now had become quite a complex network of routes.*

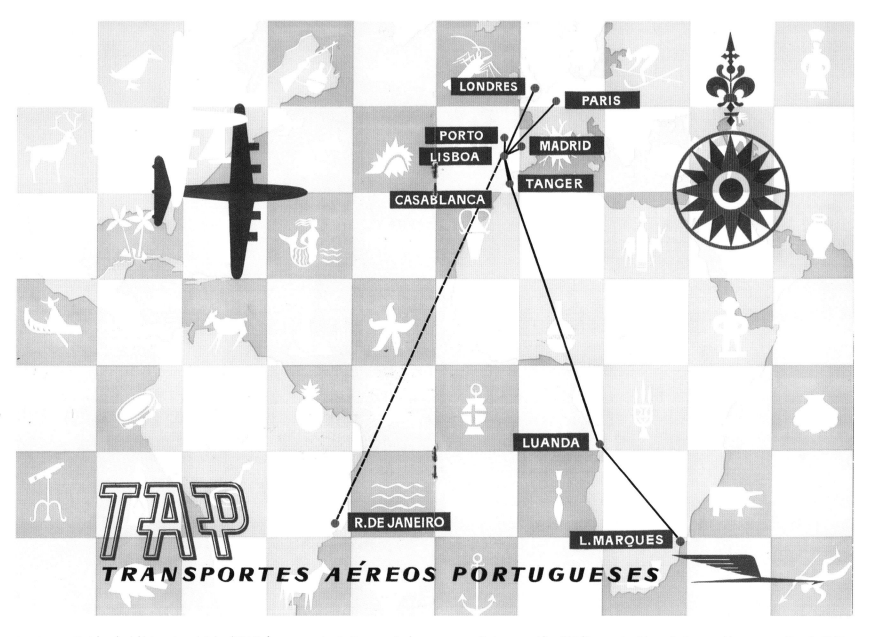

OPPOSITE LEFT: *Dutch colonial interests maintained KLM's focus on routes to Curacao, Aruba, and Central/South America, and this 1949 brochure turns the world on its side in order to fit the space (the other side twists the world the opposite way to show Australia).*

OPPOSITE RIGHT: *The Sociedad Colombo Alemana de Transportes Aéreos was formed in 1919 and merged with other lines to become Aerovías del Continente Americano S.A. (Avianca), the flag carrier for Columbia, in 1941. Avianca developed a striking visual identity, such as this early 1950s poster with checkerboard effect, a style to be repeated elsewhere.*

ABOVE: *Portuguese airline TAP (Transportes Aéreos Portugueses) began operations in 1946 with a flight between Lisbon and Madrid. It had been freed from state control a year before this checkerboard effect cover was made for the 1954 timetable. The squares also resemble tiles. Such a design was particularly appropriate, as the country prided itself on being home to one of Europe's leading ceramic artists, Maria Keil (1914–2012).*

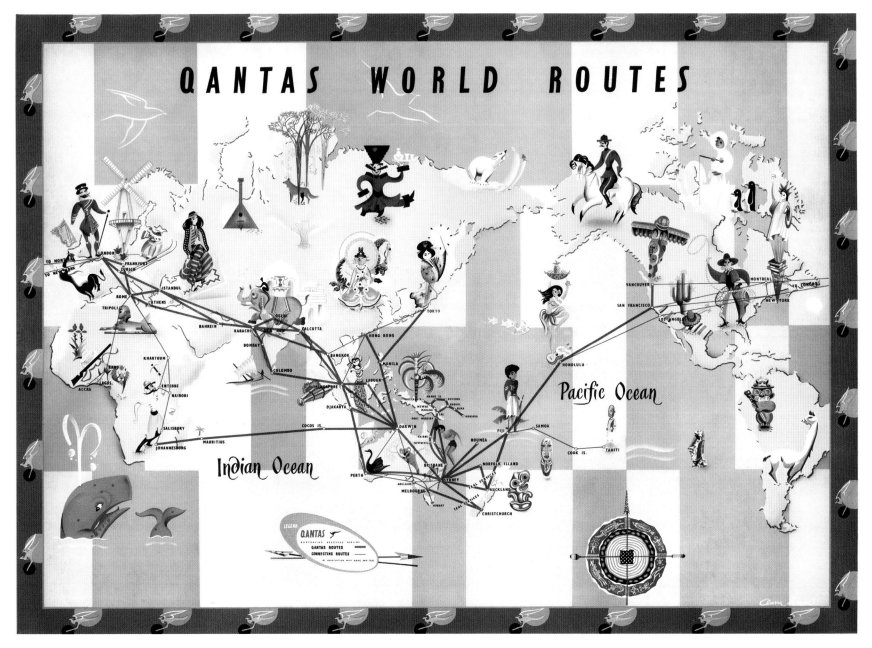

ABOVE: *A larger checkerboard effect is applied to this 1950 Qantas cartograph by Anne or Anna Drew (1916–2002), aka Joan Dent. Her whimsical Cossack and Beefeater, delicate geisha and Hawaiian Islander, and comical elephant and whale are some of the most cheerful and uplifting caricatures of any of these pictorial maps.*

OPPOSITE: *A 1949 BOAC world routes illustrated by artist E. O. Seymour. The "Eastern Hemisphere" is part of a pair; the other, not shown, covered the Americas. The contrast between informal Australian graphics and more staid British equivalents is striking.*

Radar was an another important development from the Second World War, revolutionizing air traffic control, particularly for guiding takeoffs and landings. The other legacy of the war was modern airfields, complete with long concrete runways and large hangar complexes for storage and maintenance, which were built during hostilities and no longer needed. This, alongside the increase in range and speed of the new aircraft, rendered the massive flying boats of the previous era obsolete, but traditional propeller power itself was likewise soon to be eclipsed by revolutionary technology.

Nonetheless, further refinements were still taking place, and the Douglas DC-7, introduced in 1953, was able to cruise at speeds in excess of 300 mph, carrying almost seventy passengers. This aircraft enabled American Airlines to offer the first nonstop US coast-to-coast service, although the DC-7 was ill-suited to the ambitious schedule and proved to be unreliable.

The British-built de Havilland Comet 1, the first purpose-designed civil jet airliner, pointed to a different future, in concept if not in implementation. Its sleek fuselage seated only forty passengers, initially connecting London with

 64

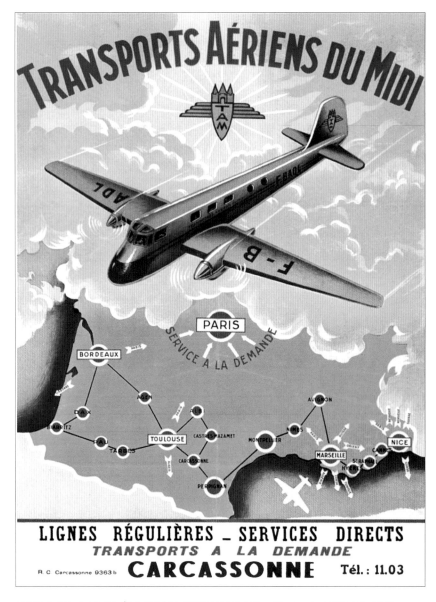

ABOVE: *Just two posters from the late 1940s French airline Transports Aériens du Midi are known to have survived, which is a shame, as they had panache—especially the London Underground–style place names.*

RIGHT: *Given French interests in the vast expanse of Africa, it is hardly surprising that so many posters were produced advertising its flights there. This late 1950s example features neatly crafted vignettes and hand-drawn sans serif lettering reminiscent of Peignot by famous poster artist A. M. Cassandre (a typeface that dates from 1937).*

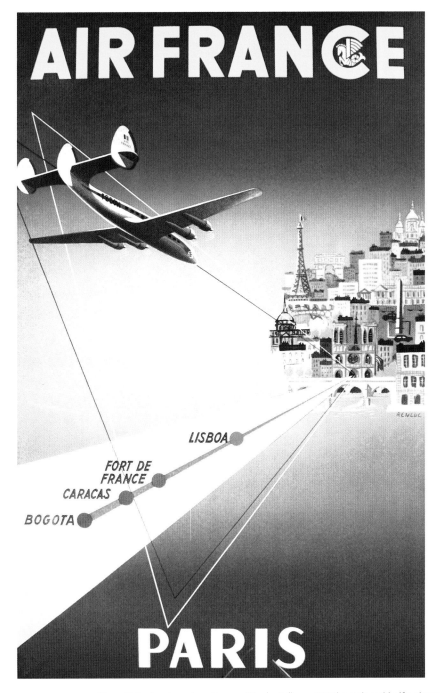

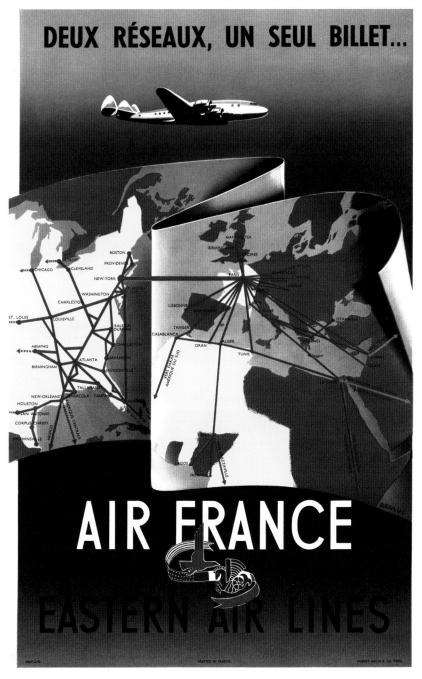

BOTH ON THIS PAGE: *The artist who signed posters as "Renluc" (born 1900) produced half a dozen for Air France. This 1953 example (ABOVE LEFT) has similarities in style to many Bernard Villemot (1911–1989) designs. Villemot had a similar knack for bunching buildings together in skyscapes, as Renluc did here. But Renluc's eye for abstraction is sharper: this play on triangles and the route to Bogotá are unique. Also unique is his treatment of the Northern Hemisphere on a 1950 poster (ABOVE RIGHT), which folds the Atlantic to imply a short crossing.*

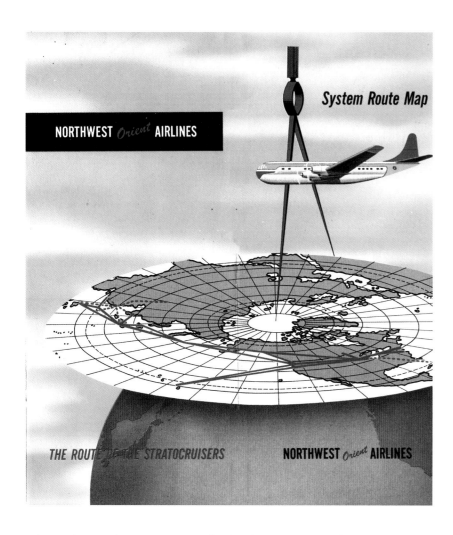

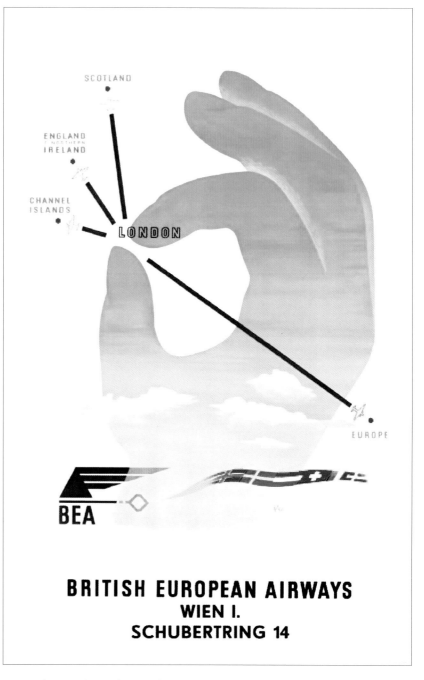

Johannesburg, Tokyo, Singapore, and Colombo. However, the fleets were soon grounded after several terrible accidents owing to poor understanding of the stresses caused by powerful engines. The true jet age would have to wait a few more years.

Worldwide, the stage was set for air travel to flourish. In the 1930s, it had been expensive and uncomfortable, and was only worthwhile for the richest of travelers. Now, increased speeds and fewer refueling stops resulted in even greater time savings, better justifying the expense for more people, and with fewer comfort sacrifices necessary. Potential passengers, as illustrated on publicity materials, are less ostentatiously wealthy than before (p. 67).

The most obvious outcome of the growing appeal of flight was the total devastation of the once-mighty US railroads. Squeezed between better aircraft and more competitive flights on one side, and improved roads and increased car ownership on the other, from 1947 to 1957, US passenger miles by train dropped from around sixty billion to less than thirty billion. The recent investments by

LEFT: *Northwest Orient Airlines produced this route map in 1956. The cover has become iconic among cartophiles, turning a new world projection into pure art.*
RIGHT: *Hans Schleger (1898–1976, pen name, Zero), another famous designer, produced this ingenious poster for BEA in 1947. He made countless ads for Britain's brands and institutions and, here, geography is stripped away, replaced by an enigmatic hand.*

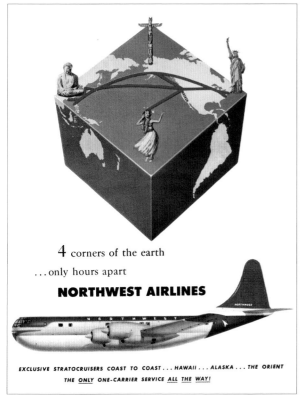

4 corners of the earth
...only hours apart

NORTHWEST AIRLINES

EXCLUSIVE STRATOCRUISERS COAST TO COAST...HAWAII...ALASKA...THE ORIENT
THE <u>ONLY</u> ONE-CARRIER SERVICE <u>ALL</u> <u>THE</u> <u>WAY</u>!

Where on *SAS* worldwide routes will you see these fascinating people?

☐ U.S.A. ☐ JAPAN ☐ GREECE ☐ ARGENTINA ☐ SCANDINAVIA
(Compare your answers with correct answers below.)

SAS
SCANDINAVIAN AIRLINES SYSTEM

THE GLOBAL AIRLINE

UNIVERSAL GENEVE – World-famous watch timing every SAS flight.

FLY TWA
JETSTREAM

POLAR
ROUTE
CALIFORNIA

LONDON · PARIS
ROME
SAN FRANCISCO
LOS ANGELES

B·O·A·C ROUTE MAP No. 1

BRITAIN · U.S.A. · CANADA · CARIBBEAN

ALL ON THIS SPREAD: *Commercial illustrators must have loved getting a commission from an airline. Not only were they the height of cool in the late 1940s and early '50s, but artistic license was positively encouraged. The depiction of the Earth that aircraft were now able to cross with ease, gave rise to a plethora of new creative ways to show the planet and entice passengers to take a voyage over!*

TOP LEFT: *Its slogan claimed "the sun never sets on Northwest Airlines," and it was keen to promote its links to the "4 corners of the earth" in 1951. The cubed globe emulated BOAC's earlier effort (p. 54).*

TOP CENTER: *Magazine ad promoting the new SAS "polar route" from 1957. It was a much quicker way between Scandinavia and the Pacific coast of North America or Japan.*

TOP RIGHT: *TWA artist David Klein (1918–2005) was quick to exploit the airline's route over the ice too, on this 1958 poster citing the "jetstream" (a name popularized by bomber pilots during World War II) and lavishly illustrating the aurora borealis*

BOTTOM LEFT: *An enigmatic flock assembled using BOAC's Speedbird logo circumnavigates the globe on this system route map cover of 1954, designed by the artist Maurice Laban (1912–1970).*

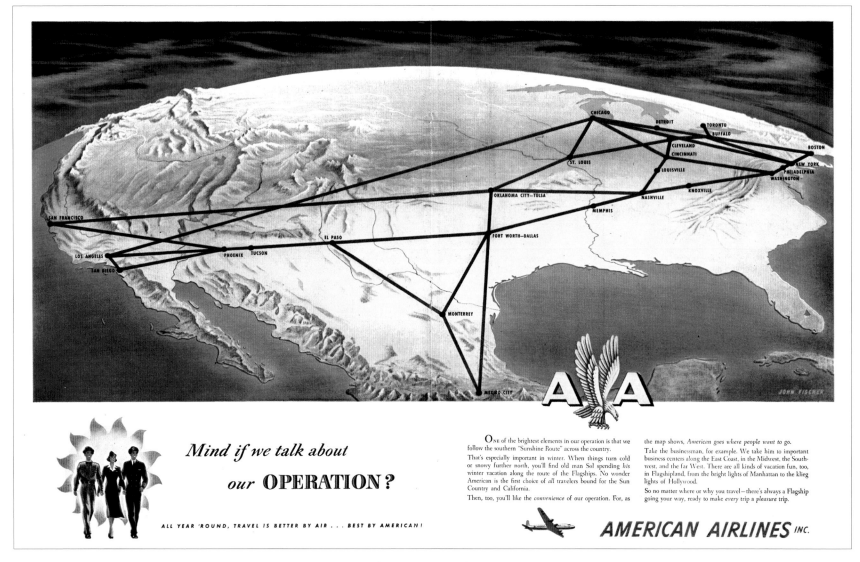

ABOVE: *Although it wouldn't be possible to see North America and the curvature of the Earth like this for another twenty years, American Airlines artist John Fischer created a good approximation, in 1949, of what the country might look like from space—and also a convincingly warm-looking ad for the airline's plentiful links to the sunnier parts of the country.*

many railroad companies into brand-new trains and station facilities was utterly wasted, and they must have subsequently sensed the futility of the battle. In the first half of the 1950s, advertising by airlines increased considerably, but US railroad advertising showed a decline.

Politically, the Second World War ushered in massive changes. Europe was war-torn, and the days of far-flung empires were coming to an end. Both the British and French retreated from most of their overseas possessions, reducing the need for government-subsidized national airlines to bind the elements together. Imperial Airways had already been renamed British Overseas Airways Corporation in 1939, and although it remained a massive operation, it could no longer rely on colonial traffic. The creation of state-owned British European Airways in 1946 emphasized that the state no longer gave priority to one

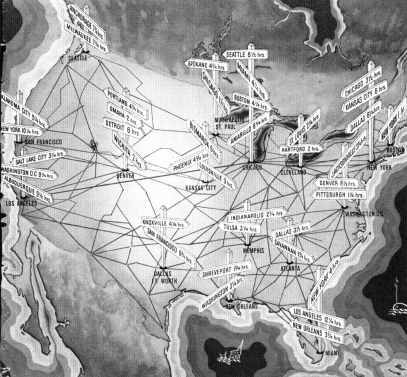

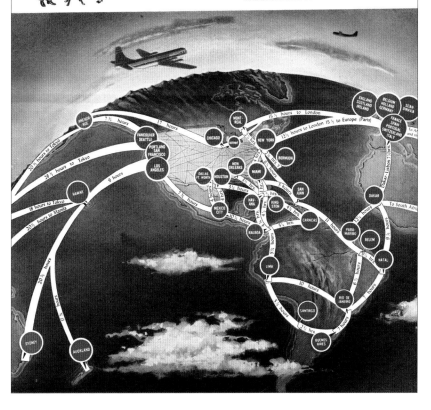

BOTH ON THIS PAGE: *United Aircraft Corporation had begun making planes as far back as 1934. By the 1950s it was specializing in Pratt & Whitney engines, propellers, and helicopters, but its approach to advertising was to publicize the general speed of air travel. Both ads show average flight times in 1950 (LEFT) and 1954 (RIGHT). The desire of designers to look down on the planet from far out in orbit was the stuff of sci-fi until the late 1960s—and what would they have made of Google Earth?*

single operator. The newly autonomous colonies would create higher-profile flag-carrying airlines to represent their countries; for example, Tata Airlines (founded in 1932) was renamed Air India in 1946, and Qantas Empire Airlines was nationalized by the Australian government, becoming merely Qantas Airways. Beyond the crumbling empires, the new world order led to other mergers and new names, such as Scandinavian Air System (SAS) in 1946.

Simultaneously, less ravaged by war, the United States expanded its operations. Before the war, only Pan Am was permitted to run international routes. Now the US government allowed Howard Hughes's Transcontinental & Western Air Transport (renamed Trans World Airlines) to join them, along with Northwest Orient (primarily serving the Pacific and Far East), Braniff Airways (for South America), and American Airlines. The new empire builders lost no time in staking

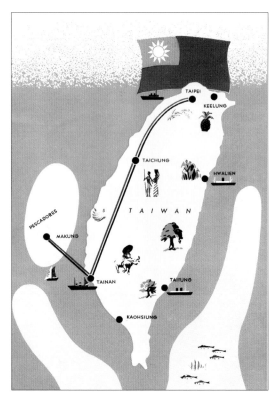

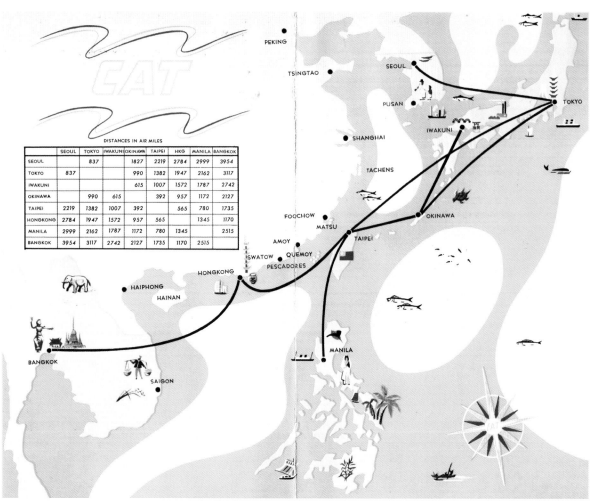

	SEOUL	TOKYO	IWAKUNI	OKINAWA	TAIPEI	HKG	MANILA	BANGKOK
SEOUL		837		1827	2219	2784	2999	3954
TOKYO	837			990	1382	1947	2162	3117
IWAKUNI				615	1007	1572	1787	2742
OKINAWA		990	615		392	957	1172	2127
TAIPEI	2219	1382	1007	392		565	780	1735
HONGKONG	2784	1947	1572	957	565		1345	1170
MANILA	2999	2162	1787	1172	780	1345		2515
BANGKOK	3954	3117	2742	2127	1735	1170	2515	

DISTANCES IN AIR MILES

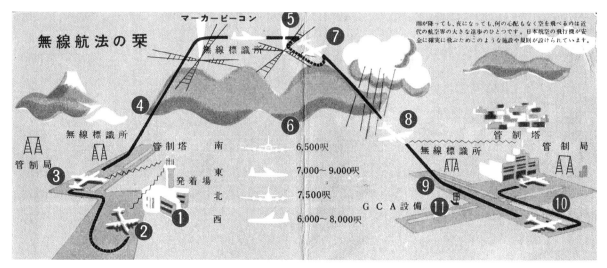

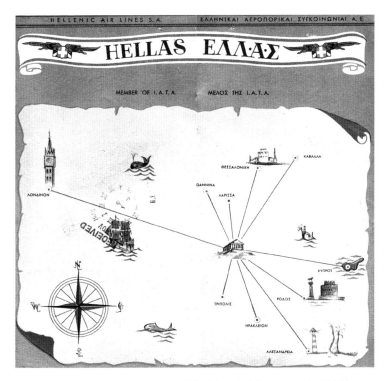

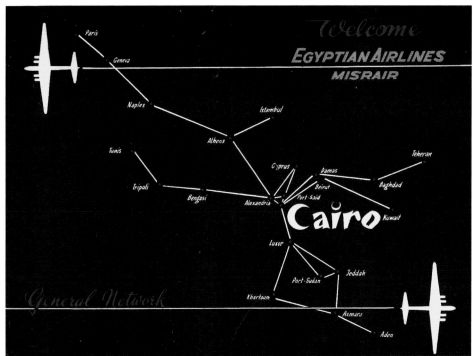

OPPOSITE TOP LEFT AND RIGHT: *CAT was a Taipei-based airline started in 1946 by Claire Lee Chennault (1890–1958) and Whiting Willauer (1906–1962) to airlift aid to the ravaged country. It was later taken over by the US government. These 1950s maps are from CAT's timetable booklet.*

OPPOSITE BOTTOM LEFT: *Thai Airways Company was founded in 1951, and this delightful poster of its original routes was produced not long after.*

OPPOSITE BOTTOM RIGHT: *Founded in 1951, Japan Air Lines (JAL) grew slowly at first. This 1956 card shows a unique schematized/pictorial flight plan, complete with airports, mountains, and altitudes.*

ABOVE LEFT: *Hellenic Air Lines started operations in 1947 and a year later produced this charming route map of services around Greece and to Egypt, Cyprus, and its longest haul service—to London. The sea is mysteriously absent here, with the Mediterranean reduced to sketchy ripples. The company ceased operating in 1951.*

ABOVE RIGHT: *Misr Airwork began in the early 1930s, and twenty years later had morphed into Misrair (aka Egyptian Airlines), a name used until 1957. This charming late 1950s network map has the air of a star constellation about it.*

BOTTOM RIGHT: *Iraqi Airways commenced operations in 1946. This 1954 timetable cover and map, centered on Baghdad, shows a limited Middle Eastern service. Its long-haul flights, using Vickers Viscount planes to London, began the next year.*

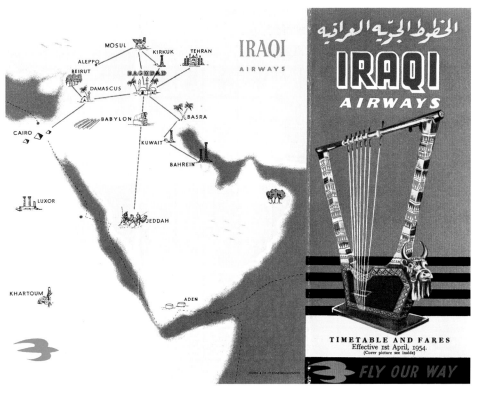

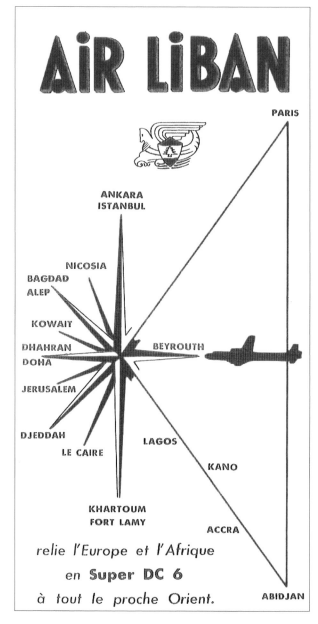

OPPOSITE LEFT: *Svitlet was a Czechoslovakian airline founded in 1948, but it was soon absorbed by CSA. This 1950 timetable cover is the only known airline map printed on the sole of a boot—Svitlet was making a pun from a folk tale about seven-league boots.*
OPPOSITE RIGHT: *In 1932 all Soviet air services were consolidated under the name Grazhdansky Vozdushny Flot, roughly translated as Civil Aviation Fleet, which became known as Aeroflot. Prior to the Second World War, its publicity had been somewhat lackluster, but this 1952 poster demonstrates its output had improved, although most of it was aimed at the tourist market.*

LEFT: *Founded in 1946 during Hungary's Soviet era, Maszovlet (renamed Malev in 1954) was quite adept at publicity. This 1952 timetable cover is a good example of its very beautiful productions during that time.*
RIGHT: *Lebanon's Compagnie Générale de Transport began in 1945 but changed its name to Air Liban in 1951. This clever play on a compass is on a timetable cover from six years later.*

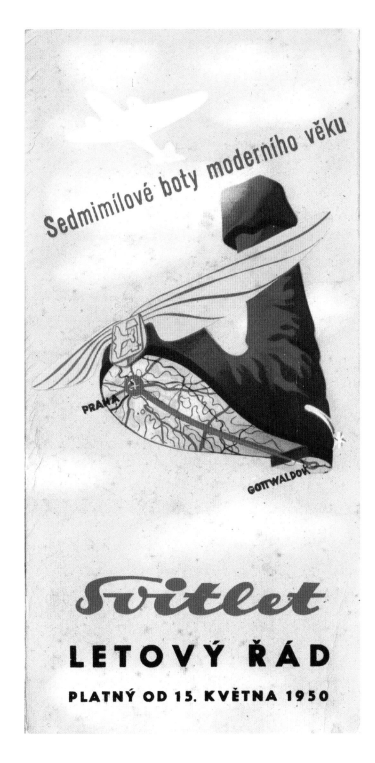

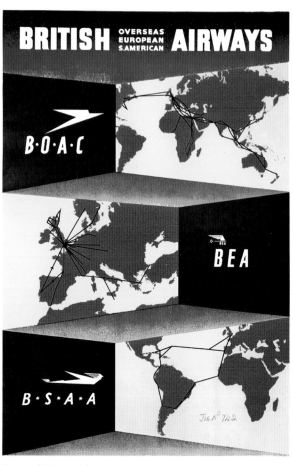

ALL ON THIS PAGE: *Three major UK operators, British South American Airways (BSAA, Latin American routes), British European Airways (BEA, mainly domestic and European services), and British Overseas Airways Corporation (BOAC, internationally focused), were all state run and worked closely together, often sharing marketing, as the 1947 joint poster implies (**RIGHT**). BOAC was the most prolific (pp. 54–55), but posters for BEA in 1946 (**CENTER** and p. 64) offered more experimental ideas. This 1948 rainbow route for BSAA (**LEFT**) was possibly the strangest.*

their global claims (pp. 56–57) with assured maps that in no way implied that they were relative newcomers.

Many commentators call the 1950s a golden age of passenger flight. The decade certainly ended on a high, with the rehabilitated de Havilland Comet and the Boeing 707 beginning transatlantic flights in 1958, but the year also seems to have been a turning point. Until then, there was a straightforward evolution of technology first developed in the 1930s. Relatively small, slow propeller aircraft plied the world's skies, still needing refueling stops if the great oceans were to be crossed. They had reached the pinnacle of their development.

Air travel was still expensive but worthwhile for people who needed to cross the Atlantic Ocean, or the United States, by the fastest available option. Comfort was improving, and some of the glamor of flight remained. Furthermore, costs were coming down, and profitability was increasing. The airlines were poised to

win the final battle against their rivals—the seagoing ocean liners and the railroads—but one more piece of the jigsaw was still required, then they would be able to capture the markets for long-distance travel.

Graphically, reflecting massive post-war changes, map imagery was at a crossroads. Sumptuous Art Deco imagery evoked pre-war glamor and excitement (pp. 62–63, 71), and an elegantly dressed woman makes the same point more subtly (p. 58). Gorgeous global pictorial maps depict exotic destinations that are closer than they had ever been before (pp. 52–53, 56, 60–61, 77), many of these designs clearly inspired by Lucien Boucher's work for Air France. Illustrated embellishment was even added to a flight departure time diagram in this era (p. 78). In some cases, however, the images on these maps are cartoonlike, almost childish, pointing toward a desire for mass-market appeal (compare the pair of maps on pp. 60–61). In contrast to the pictorial maps, the

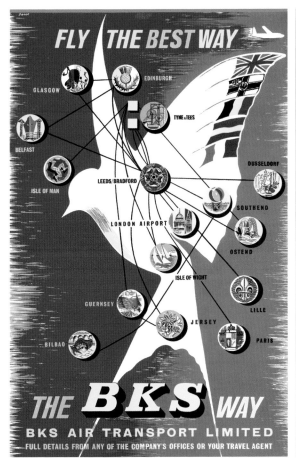

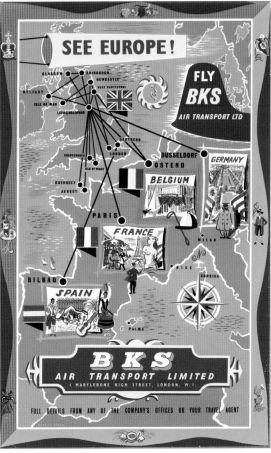

LEFT AND CENTER: *BKS was a small independent UK company founded as BKS Aero Charter in 1951. These posters were issued around 1954, after it had gained licenses to run some limited scheduled services, altered its name, and became based out of Newcastle-upon-Tyne. Maps were bright and colorful, aimed at casual, short-hop, occasional visitors.*

ABOVE: *BKS was perhaps being directly targeted by BEA in this 1953 flowery offering, using bright and jolly posters to advertise short-hop destinations.*

plainer ones (p. 57) are no less dramatic, although astrological imagery is difficult to reconcile with modern flight. At the other extreme, road signs seem to suggest that air travel might be almost as mundane as an automobile trip (pp. 67, 72).

There are many other graphical clues pointing toward a desire to expand the market for air travel. The continuing rise of magazine advertisements signals this (pp. 50, 65–67, 74–75), and several depictions of people in far-off lands seem intended to suggest that friendly natives will be delighted to greet nervous travelers (pp. 50, 65, 68). Not surprisingly, given the recent war, military precision of operations also becomes a theme (pp. 66, 74–75), again perhaps to reassure would-be travelers that aircraft operations are rigorously organized. This era was also an all-too-short-lived golden age for commercial artistry, beautifully drawn and colored-in machines and people grace in which lowly mass-market

publications. Elsewhere, more creative approaches to design display an abstract playfulness, folding, slicing, and reshaping the globe in entertaining ways (pp. 54–55, 63–65), although placing the landscape underneath a boot is probably an unfortunate choice (p. 71). There is also a brasher, more colorful style emerging, which points toward the 1960s (p. 73).

Abstract diagrams were first seen in the 1930s, in tune with modernism's principle of fit-for-purpose simplicity alongside avoidance of unnecessary embellishment (pp. 35–37). Diagrams also make further appearances in this era, with landscapes stripped away, leaving just the simplest of depictions (pp. 63, 64, 69, 70, 76). However, these are isolated examples and, similarly for transit maps showing urban rail networks, schematized representations did not become commonplace until the 1960s. With such a range of cartography, 1946 to 1957 was certainly a golden age for variety in the world of airline maps.

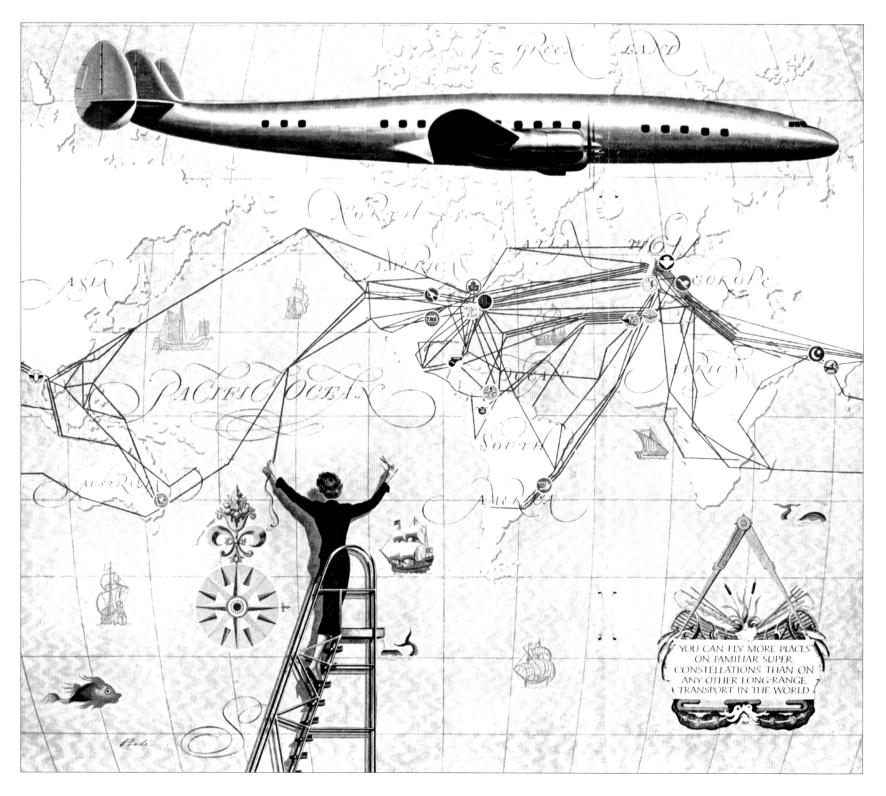

ALL ON THIS SPREAD: *A fun quirk of some publicity around airline cartography is what the authors describe as "giant people pointing at maps" or perhaps "Bond villains planning world domination." These examples include a late 1940s Canadian Pacific Airlines poster (**ABOVE LEFT**) with a gargantuan stewardess about to bowl the Earth down some galactic alley; a 1952 magazine ad from United Air Lines (**ABOVE RIGHT**)—foreseeing the look of some twenty-first-century plane-spotting apps, or planning some Machiavellian air raid on Middle America? Then there is a stunningly imagined map wall from a 1955 Lockheed press ad (**OPPOSITE**) showing the huge number of air routes across the planet that use its aircraft, while the decorator balances on a slightly precarious-looking stepladder.*

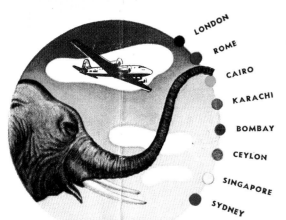

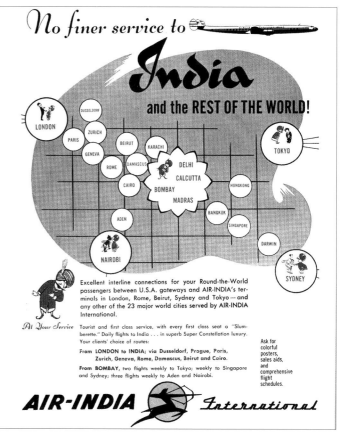

TOP: *BOAC took great pride publicizing the Speedbirds for long-haul flights, as on this delightful 1950s map.*
BOTTOM LEFT: *An early 1950s Air Ceylon press advertisement with a wonderfully simplified route diagram.*
BOTTOM RIGHT: *Magazine ad from 1957 showing Air India International destinations in an unusual format.*

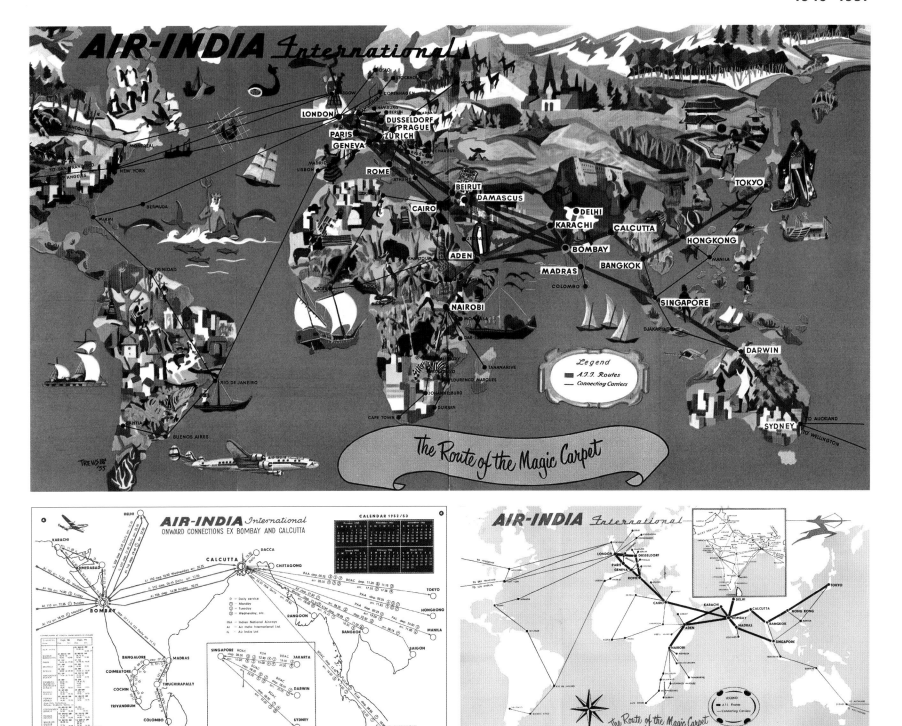

ALL ON THIS PAGE: *In 1932 the Tata Sons set up India's first commercial airline run by local people. It became Air India Limited in 1946, with "International" added two years later. Early maps were timetabled like the 1952 version (**BOTTOM LEFT**); three years later (**BOTTOM RIGHT**), graphics were subdued, but the step change happened when color was embraced by 1955 (**TOP**).*

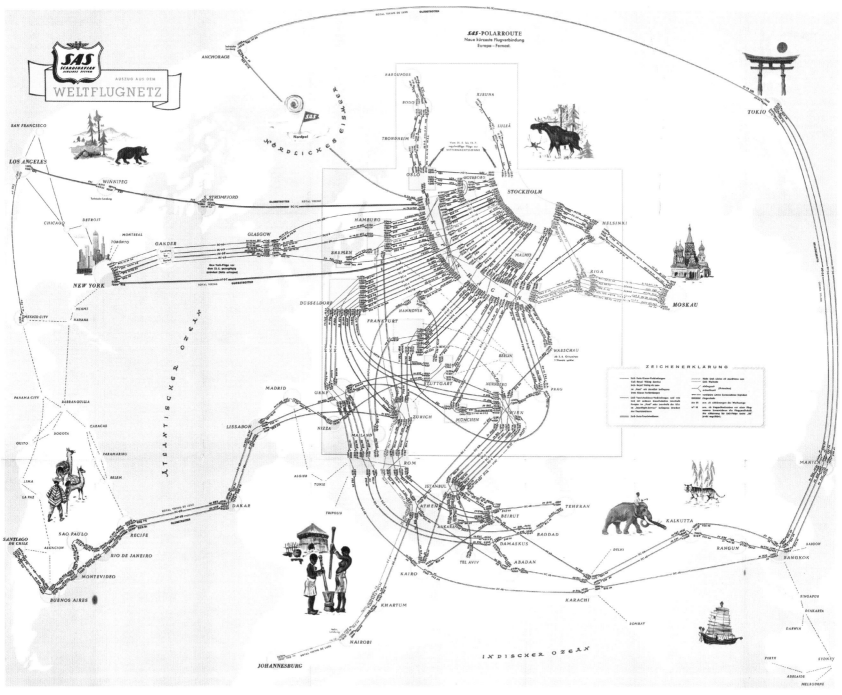

ABOVE: *Using a mesmerizing mix of departure and arrival times, plane makes, and flight numbers, this 1950s SAS fold-out map attempts to enliven what would otherwise be a dry chart. It feels like a mélange between styles of previous diagrams (pp. 44–45), and the color palette, vignettes, and even the kooky projection are unique to this SAS publication.*

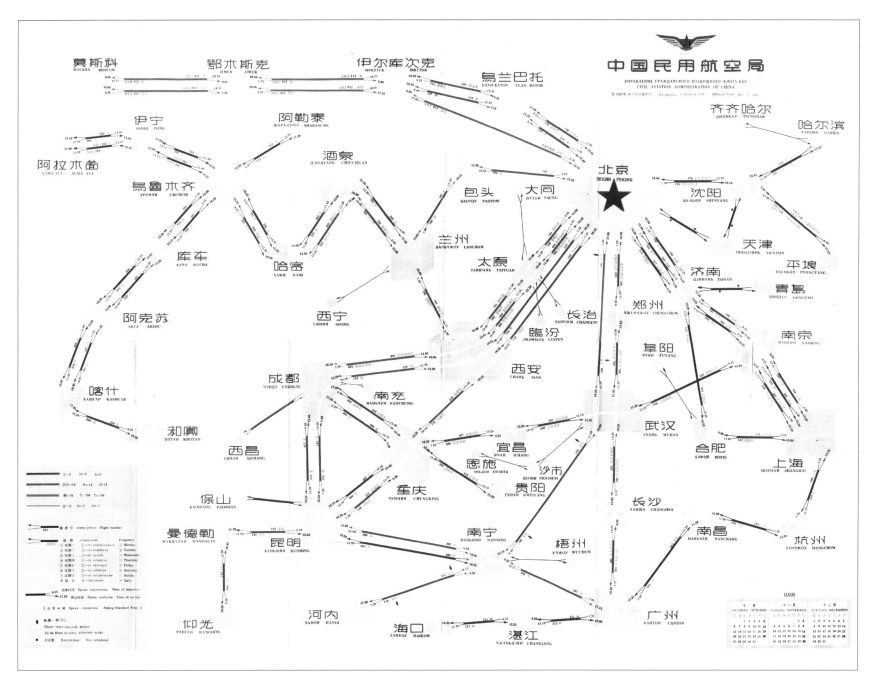

ABOVE: *The Swedes were not the only ones making use of complex service-pattern diagrams. This 1958 Civil Aviation Administration of China (CAAC) map also shows arrival and departure times to each destination, plane types and flight numbers, and the number of flights per week. What the uneven shapes in each location represent is currently a mystery. Airport land?*

NEWEST. FASTEST AIRLINER IN THE WORLD!

B·O·A·C *Comet* **JETLINER**

Terminal 4: Jets shrink the Earth

Civil aviation was slow to adopt jet technology. The year 1958 saw the first transatlantic jet-powered passenger flight, but once the benefits of increased speed and altitude were fully appreciated, rapid introduction followed. The excitement over the frontiers in air travel were reflected in new maps, with bold, almost psychedelic colors. Times were changing in graphic design, however, and the gorgeous commercial artistry of previous decades was in rapid decline, partly replaced by a more cartoon-like, less sophisticated style.

BACKGROUND: *A large-format Pan Am poster from 1964 with somewhat exaggerated mountain range sizes.*
LEFT: *Although BOAC was the first airline to use jets in public service (from 1952), and advertised the de Havilland Comet with this cutaway, it was an unwise image; they were grounded by 1954 after a spate of accidents in which they disintegrated at altitude.*
OPPOSITE: *Ingenious "Spiral Polar Projection" for SAS on this 1960 poster—but sadly the creator is not named. Technologically, it gives the image of looking forward enthusiastically to the jet age, whereas from a design point of view, gloriously looking back to styles depicted in the previous chapter. In its scope and execution, this powerful image represents something of a design pinnacle.*

CREATED ESPECIALLY FOR SCANDINAVIAN AIRLINES SYSTEM TO ILLUSTRATE ITS WORLDWIDE ROUTES

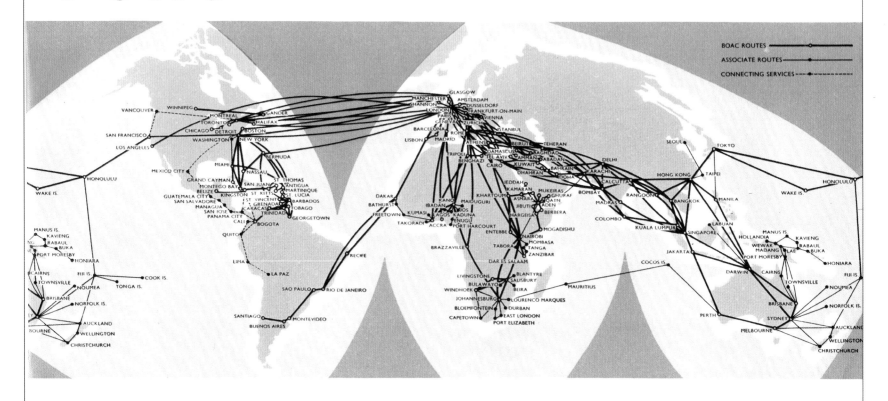

The B.O.A.C. Jetliner Fleet of Rolls-Royce 707s, Comets and Britannias operates an intercontinental route network, radiating from Europe to all parts of the world. Frequent BEA flights within continental Europe connect with B.O.A.C.'s world-wide jet services at London, Manchester, Glasgow, Zurich, Geneva, Frankfurt, Madrid, Lisbon, Rome and other European cities.

BRITISH OVERSEAS AIRWAYS CORPORATION

BEA are General Sales Agents for B.O.A.C. in Europe

ABOVE: *BOAC was able to reintroduce Comet jets in 1958 once their safety issues were rectified, starting the first transatlantic jet service. By the end of the 1950s, just as the new decade dawned, BOAC was in its prime, operating one of the biggest jet-propelled fleets in the world, as this busy map indicates.*

BOAC commenced the first transatlantic passenger jet service in October 1958, using a British de Havilland Comet version 4, newly redesigned to cure earlier catastrophic metal fatigue problems, and to carry more fuel and passengers— eighty-four versus forty—faster than ever before. With a cruising speed of over 550 mph, crossing from London to New York required over ten hours westbound (with refueling in Newfoundland), but the eastbound trip could be completed in under seven hours, owing to prevailing atmospheric conditions. Later that month, British aviation pride was dealt a double blow: the well-documented limitations of the Comet spurred rival builders to create better aircraft, and Pan Am introduced transatlantic Boeing 707 flights. This airliner carried almost one hundred more

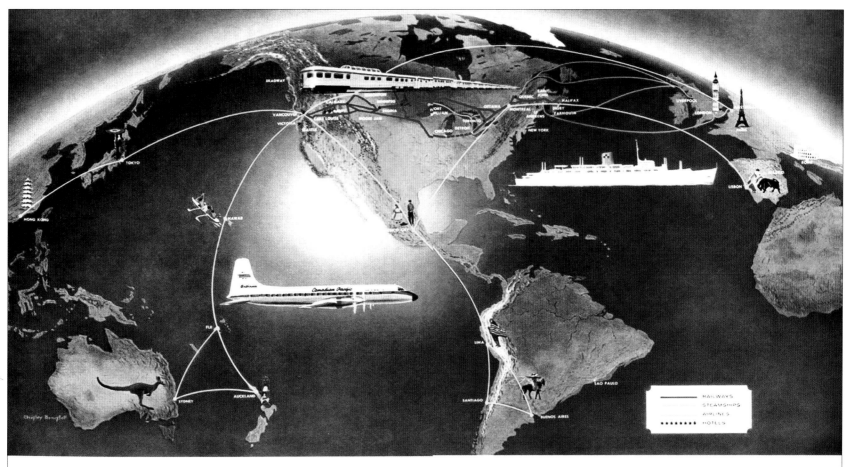

By land, by sea, by air, Canadian Pacific — the world's most complete transportation system — serves five continents

Daily, Canadian Pacific's sleek scenic dome streamliner train, "The Canadian," speeds across Canada on the Banff-Lake Louise route.

Graceful White Empress ships glide majestically down the St. Lawrence River past picturesque French Canadian farms and villages.

Jet-prop Britannias and DC6-Bs take wing regularly to five continents, and Canadian Pacific also provides air service across Canada.

Night and day, commerce moves by Canadian Pacific coordinated rail and trucking services, by ocean, inland and coastal ships.

Across Canada, Canadian Pacific operates a chain of hospitable hotels typified by baronial Banff Springs and Chateau Lake Louise in the Canadian Rockies, and the Royal York in Toronto, largest hotel in the British Commonwealth.

And rounding out its many services, Canadian Pacific maintains a system of world-wide communications and express services.

This is Canadian Pacific — the

World's Most Complete Transportation System — operating and expanding more than 85,000 route miles by land, sea and air.

Canadian Pacific

The World's Most Complete Transportation System

RAILWAYS • STEAMSHIPS • AIRLINES
HOTELS • EXPRESS • COMMUNICATIONS
TRUCKING • PIGGYBACK

ABOVE: *Formed in 1942 by the great namesake railway company, Canadian Pacific Airlines was delivering over 500 million passenger miles on scheduled flights by 1960. This map of a year earlier partly explains why: it was able to offer rail, sea, and air connectivity from Canada right across the world under "the world's most complete transportation system."*

passengers and was faster with a longer range, as was the Douglas DC-8 that was introduced soon afterward. The benefits of the new aircraft proved irresistible to operators; international flights were soon offered globally, and hundreds of jets were sold. Jet engines were more reliable and easier to maintain than the piston equivalents, and their increased power and speed gave them a competitive edge.

They could fly over the weather, and the poles if necessary, taking the fastest route possible from point A to point B. The downside was that they consumed more fuel, but because they were more powerful, they could carry more than enough to compensate for this. Early jets proved far more cost-effective on longer routes, but the French-designed Sud-Est Aviation Caravelle, introduced in 1959, showed that

 93

ABOVE: *Inter-Island Airways, which began in 1929, changed its name to Hawaiian Airlines in 1941, but was still running old Douglas DC-3 propeller planes and Sikorsky seaplanes until it entered the jet age in 1966. Perhaps its ties to older technology are reflected in this beautiful, but slightly dated-looking map from 1960? Or, maybe, Hawaiians were keen on nostalgia, as the Aloha map (p. 108) likewise seems to imply. Either way, the coloring and style are as bright and attractive as the shirts that bear the islands' name.*

OPPOSITE: *Air India International had already made the move to richly illustrated color maps in the 1950s (p. 77), but its introduction of jets in the 1960s (Boeing 707-420s) seemed to spur an already highly competent publicity department to even greater achievements. In 1962, for example, as the company truncated its name to simply Air India, it became the world's first all-jet plane airline. Perhaps looking forward toward brighter times, in that year it produced both a pink and blue sea map.*

AIR-INDIA
ROUTE MAP

AIR-INDIA
ROUTE MAP

LEFT: *New Zealand National Airways Corporation (NAC) was founded in 1947 by a state-led merger of several operations. It moved rapidly from providing simply domestic services to adding nearby international destinations around the South Pacific, but it never fully expanded to long-haul routes (which were catered for by Air New Zealand, with whom it merged in 1978). This charming pictorial map, with happy smiling vignettes, dates from the mid-1960s when it was still operating turboprop Viscount planes.*

TOP CENTER: *El Al operated its first flight as Israel's national airline in 1949 (Tel Aviv to Rome and Paris). International services began a year later, with New York and Johannesburg on offer. With the arrival of Bristol Britannia turboprops in 1957, it ran a famous ad campaign, "No Goose—No Gander," because its planes no longer required refueling stops at Gander Airport in Goose Bay, northern Canada. Boeing 707 jets were introduced in 1961, kicking off the world's then longest nonstop commercial route (Tel Aviv to New York). This whimsically illustrated map from that era commemorates various battles across Europe, though by modern standards the pointing of tanks at Tunisians would clearly be unacceptable.*

אל על טסה מעל שבילי ההסטוריה

AL FLIES ACROSS THE PATHS OF HISTORY
OLE LES GRANDES ROUTES DE L'HISTOIRE

RIGHT: *KLM had introduced turboprops on many European routes in 1957 (Vickers Viscount 800s), and Douglas DC-8 jets began operating in 1960, the year of this map (which oddly shows an earlier four-engine Lockheed in the corner). Like the other maps on this spread, the KLM one also features vignettes of local attractions, though these have moved away from people (just two shown) and on to landmarks or regional culinary delicacies.*

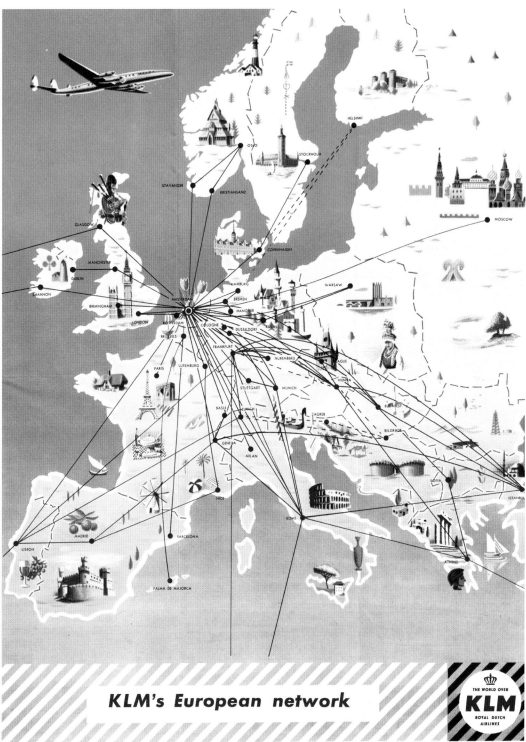

KLM's European network

THE WORLD OVER
KLM
ROYAL DUTCH
AIRLINES

ABOVE: *By the time Lucien Boucher completed this 1959 Air France cartograph, he had developed his own projection (a variation on Azimuthal equidistant) to show the planet with France at center point, and Canada, Alaska, and Siberia curving around the North Pole. These intricate designs had started to be known as "planispheres." Air France was about to go through major changes: domestic routes were transferred to Air Inter (1960, with Air France owning a stake), and Air Afrique and Union Aéromaritime de Transport were also created.*

DEUTSCHE LUFTHANSA DEUTSCHE DEMOKRATISCHE REPUBLIK

DIE WICHTIGSTEN FLUGSTRECKEN DER DEUTSCHEN LUFTHANSA UND IHRER VERTRAGSPARTNER

ABOVE: *A giant four-engine jet casts a shadow over the Earth alongside gargantuan landmarks in this 1962 pocket fold-out route map from Lufthansa. Interestingly, the only other forms of transport shown are sailboats, a freighter, and an oceangoing liner. Were the designers trying to instill the benefits of fast jet travel over shipping? The tagline reads: "The most important air routes of Lufthansa and its contractual partners."*

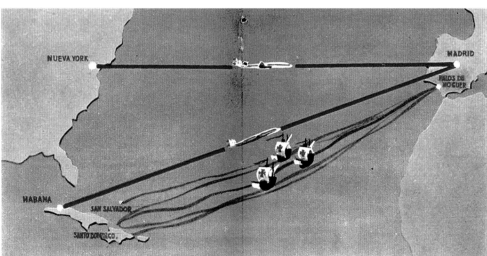

ALL ON THIS PAGE: *Iberia can trace its history to the Compañía Aérea de Transportes in 1927, which flew between Madrid and Barcelona. Following several mergers and state intervention, it was nationalized in 1944, becoming the first company to fly between Europe and South America after the war. The Spanish government allowed Avicao (p. 97) to run domestic routes so that Iberia became internationally focused. The Montreal Convention on International Civil Aviation (1954) kick-started the tourist and vacation industry by air. It led to a quadrupling of passengers in the 1950s. Two late 1950s maps (**TOP LEFT AND BOTTOM LEFT**) contrast air travel with that by sea. An early 1960s advertisement (**TOP RIGHT**) shows innovation that would influence future Iberia styles (p. 133).*

ALL OPPOSITE: *The early 1960s saw big changes for Aeroflot; its destinations increased to over one hundred from Moscow alone, and it introduced the Tupolev Tu-114—a massive, fast, long-range turboprop—and the Tu-124 jet for domestic routes. The map (**TOP**) featuring vignettes for some larger cities illustrates the extent of its network in 1962. Also from 1962, a more colorful example (**BOTTOM RIGHT**) replaces vignettes with red stars. The country is so vast that regional maps such as the one centered on Novosibirsk, Central Asia, were also common (**BOTTOM LEFT**, mid-1960s).*

ВОЗДУШНЫЕ ЛИНИИ

ВОЗДУШНЫЕ ЛИНИИ ЗАПАДНОЙ СИБИРИ

ВОЗДУШНЫЕ ЛИНИИ СССР

АЭРОФЛОТ

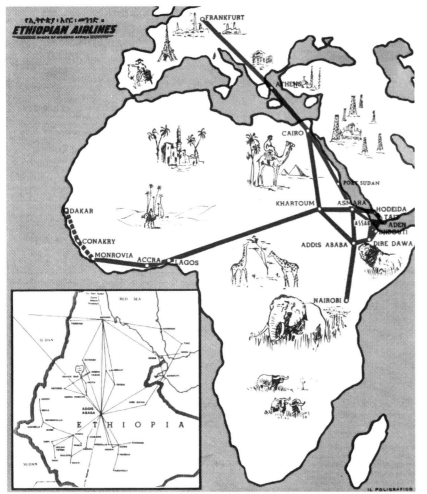

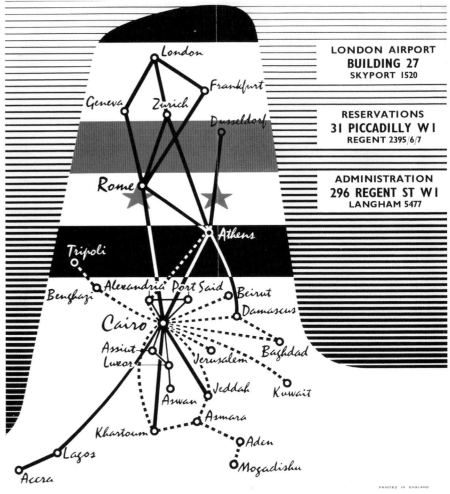

TOP LEFT: *Ethiopian Airlines began domestic flights in 1946, with international ones starting in 1951. This map was produced in 1961 on the eve of Ethiopia entering the jet age: Boeing 720s arrived in 1962.*

TOP RIGHT: *From 1957 (until the 1970s) Egyptian Airlines was known as United Arab Airlines. This early 1960s map includes a stylized tail painted in the flag of the newly formed United Arab Republic.*

BOTTOM CENTER: *Formed in 1946, Sudan Airways is one of Africa's oldest airlines. The poster designed by John Cottrell is one of a pair he produced for the airline in the mid-1960s.*

BOTTOM RIGHT: *Société Tunisienne de l'Air was formed in 1948 by the Tunisian government, with financial input and stakeholdings from Air France. It began using jet aircraft in 1961, and this colorful timetable cover dates from 1965.*

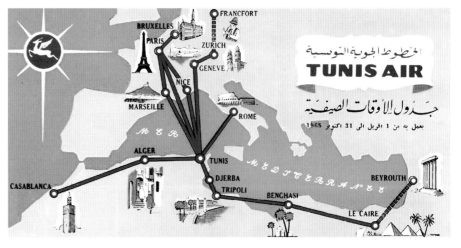

MIDDLE EAST AIRLINES *Routes* EUROPE-THE MIDDLE EAST-PAKISTAN-INDIA

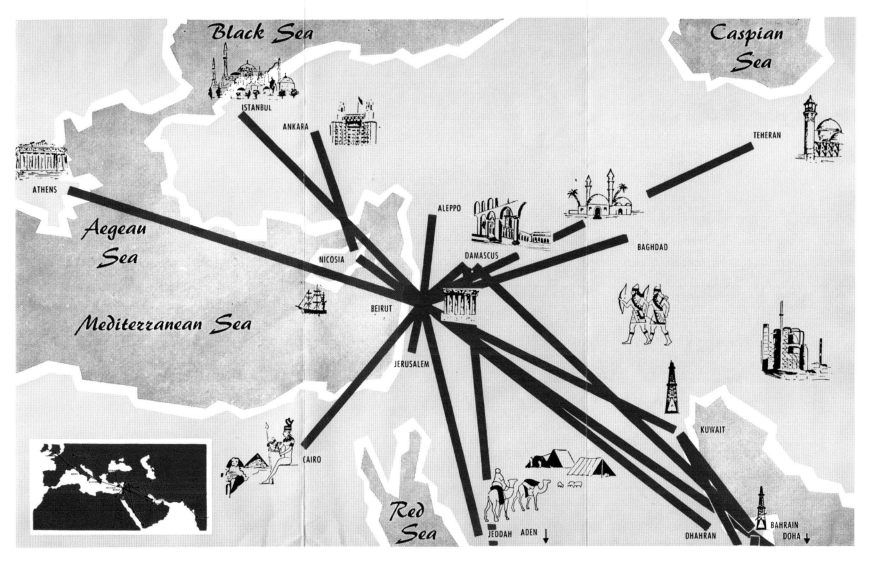

ABOVE: *Beginning in 1946, with flights between Beirut and Nicosia, Middle Eastern Airlines had operational support from BOAC. Pan Am temporarily took a commanding stake but was replaced again by BOAC in 1955. This only lasted until the early 1960s, when this map was produced, by which time MEA had merged with Lebanese International Airways.*

jets for short-haul flights could also be economic. This spurred the development of the Boeing 727, introduced in 1964.

From the passenger point of view, jets were quieter than propeller aircraft and less prone to vibration, but there was one downside: the rise of jet lag as more and more people traveled faster and farther. The physiological problems of crossing several time zones within a few hours upset body rhythms, causing disorientation and insomnia.

The battle for long distance travel was short. For the transatlantic trip, less than twelve hours by jet aircraft versus around seven days by ocean liner, meant that victory was almost brutal: by the early 1960s just 5 percent of transatlantic

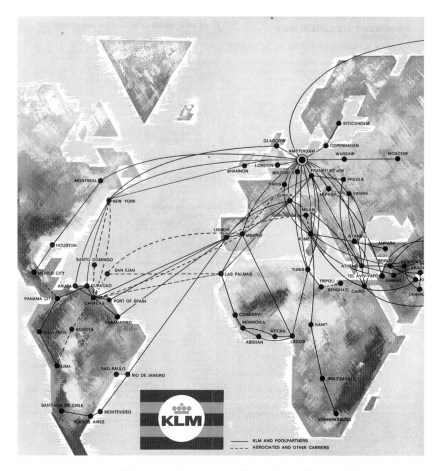

ALL ON THIS SPREAD: *New ideas sometimes pop up all over the world at roughly the same time. Do artists see inspiration in each other's work, or is it just a feeling in the ether? Zeitgeist suddenly finds bursts of triangles in airline maps during the 1960s.*
TOP LEFT: *An abstract treatment of the globe with triangular folds, by an unknown artist, makes a bright cover for this 1963 CSA (Czechoslovak State Airlines) timetable.*
TOP RIGHT: *Greenland, Scandinavia, Brittany, and Hudson Bay all get a very angular feel on this 1966 map of KLM routes.*
BOTTOM LEFT: *Artist Guy Georget (1911–1992) puts isosceles in the ascendancy on a 1962 Air Afrique poster. Note the African cities are simply marked, not linked by air route lines.*

travel was by sea. In January 1959, American Airlines offered the first US transcontinental jet service, taking just five hours, compared with eight by propeller aircraft. With a journey of around three days by train, it is no surprise that by the end of the 1960s, less than 10 percent of long-distance US journeys were by railroad. Air travel was now relatively less expensive than in previous eras, although by no means cheap, but time and convenience more than made up for costs. From just over 100 million air journeys per year in 1958, the jet-age destruction of long-distance travel by sea and railroad saw this figure almost double.

ABOVE: *Air Afrique had been a state-owned operation covering several francophone West African states. When it was spun out from Air France in 1961, artist M. Bourie (about whom little is known) was commissioned to produce this stained glass effect poster. Although no exact routes are shown, the idea was that the African continent is well connected to the entire world.*

Despite the engineering revolution, which made airliners more economical to operate and more accessible to passengers than ever before, this era was surprisingly stable in terms of operators. Of the maps in this chapter, only the Air Afrique brand was new, founded in 1961. Every single other operator had been offering passenger flights since the 1940s. Although the market was expanding, the expense of entering it was prohibitive at the time, and there also would have been the resistance of governments to allowing their flag carriers to be challenged by upstarts.

Not surprisingly, jet aircraft took their places on maps as soon as airlines purchased them. Also noticeable is the steady shrinkage in the size of the aircraft embellishments. In the 1930s, they dominated the designs, but in the 1960s, their use was far more discreet. Also noteworthy is the decline in commercial artistry. The beautiful illustrations that graced magazine advertisements previously were now being replaced by simpler graphics. Some of the ones that filled the vacuum—cartoonlike illustrations—are almost infantile (pp. 86–87, 101). This is the decade when Lucien Boucher, creator of the spectacular maps for Air France (p. 88), put down his tools for the last time. Although the era of gorgeous, artistic pictorial maps was drawing to a close, the concept is still very much in evidence (pp. 83–85, 89, 99). The almost somber elegance that characterized posters in the 1950s was also fading in popularity, with a few remaining, looking out of place (p. 100).

TOP LEFT: *Having commenced operations in 1948, Aviaco built a network of secondary routes independently of Iberia. By the time of this colorful 1966 map, the two companies were coordinating operations.*

TOP RIGHT: *A clever 1963 montage of snippets from Air France's early 1960s range of tourist leaflets, with larger illustrations by Guy Georget.*

OPPOSITE: *Delta began adding jets, such as the DC-8 and Convair 880 (shown silhouetted at the top), to its routes in 1960, the same year this beautiful map was produced.*

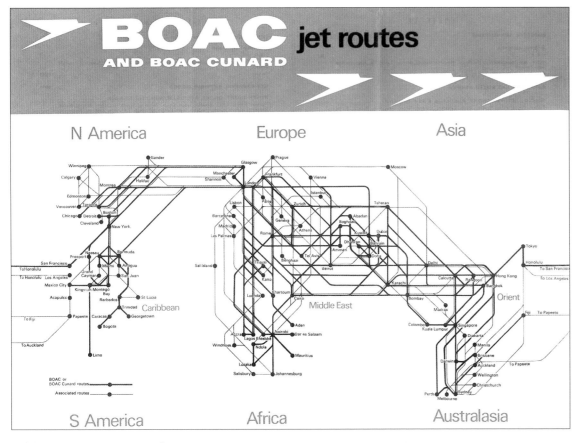

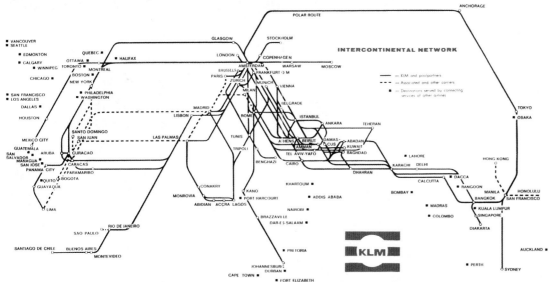

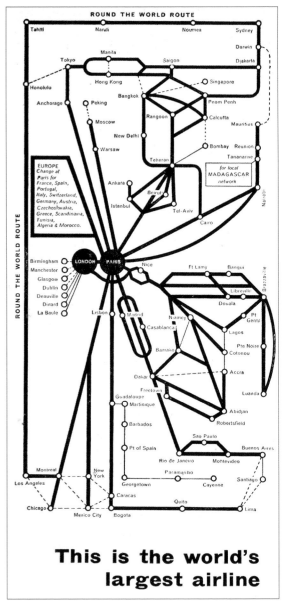

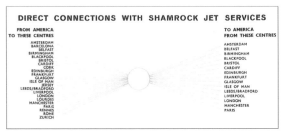

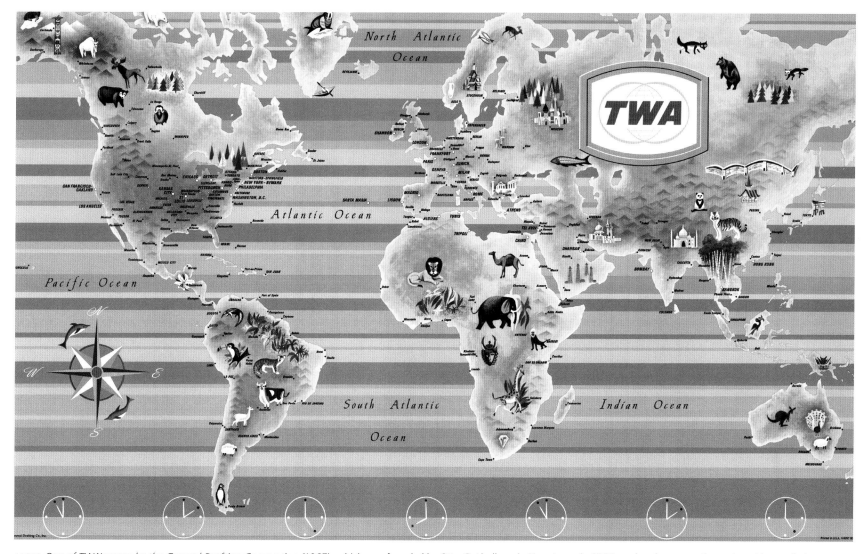

ABOVE: *One of TWA's maps by the General Drafting Corporation (1967), which was founded by Otto G. Lindberg in New Jersey in 1909, and ranks among the major mid-twentieth-century cartographers (another was Rand McNally). Note that the vignettes have moved on from indigenous people to wildlife and landmarks. Also note that this cartograph is uncluttered by the lines of the routes themselves; just TWA's destinations are shown. With no visible links, however, how could the user differentiate between direct and connecting services?*

ALL OPPOSITE: *Although used for decades (especially in the 1930s), diagrammatic airline route maps had largely slipped out of favor. The 1960s, however, saw a resurgence, and the concept would be popular in every subsequent decade.*
TOP LEFT: *BOAC's use of a 45-degree diagonal-dominated schematic in 1966. Although the change was not pursued by BOAC, it was for its descendant, British Airways (p. 130).*
BOTTOM LEFT: *KLM's 60-degree diagonal-based world map from 1964.*
TOP RIGHT: *Air France was experimenting with multiple angles in 1960 on a schematic design, but this style was not repeated.*
BOTTOM RIGHT: *One of the more abstract schematics, produced for Aer Lingus, early 1960s.*

That is not to say these times were dull. The cartoon graphics were obviously meant to promote a friendlier image, and many of the service maps are attractive in their own ways (pp. 91–93). The world became very colorful in this era, with almost psychedelic combinations resulting in unique vibrancy (pp. 96–97). Putting together geometric shapes with colors resulted in some splendid examples (pp. 94–95). Abstract diagrams also returned in force in this decade. Last seen in quantity in the 1930s, presumably these were an attempt to help users understand the complicated networks, although route color coding might have been useful, as per the transit maps that they were impersonating (p. 98).

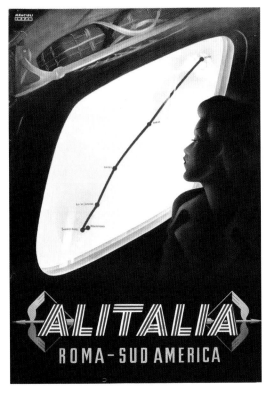

TOP AND BOTTOM LEFT: *Good ideas are timeless: 1950s Alitalia and 1962 BOAC—though the planes would have to have been in orbit to see this much of the planet!*
RIGHT: *Polish born Jan Le Witt (1907–1991) and George Him (1900–1982) were a design partnership behind scores of memorable images, like this 1960 poster for Pan Am.*

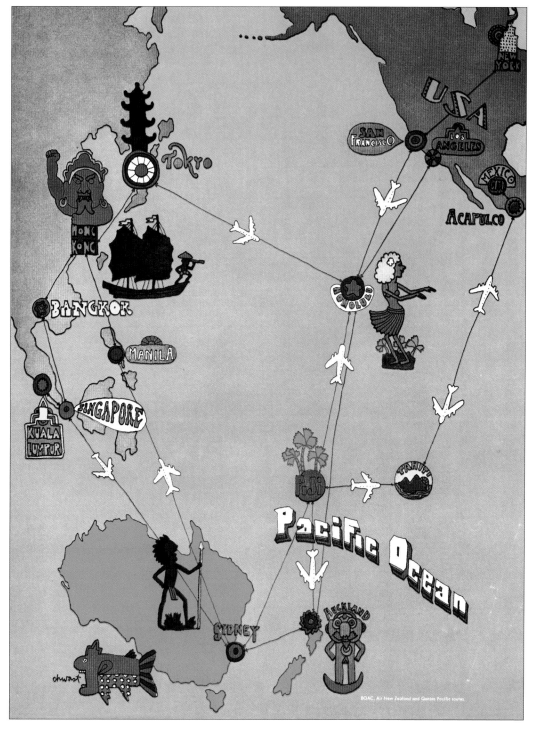

ALL: *The '60s were a decade of colorful, playful advertisements: Panagra, 1960* (**TOP LEFT**)*; Lufthansa, 1964* (**BOTTOM LEFT**)*; and BOAC, 1967* (**RIGHT**)*.*

Terminal 5: Wide-bodied world

The economics of air flight demanded more capacity, and the first prototype Boeing 747 jumbo jet rolled out in 1968. The 1973 fuel crisis ended the short-lived optimism, however. Colorful 1960s-inspired maps were still in evidence, but bland tones began to take over. The rise of photography filled the rest of the vacuum that had been created by the demise of commercial artistry. Abstract diagrams were also in the ascendency, with a wealth of simplification techniques attempted, presumably in the hope that the vast networks could be made more accessible to the public.

BACKGROUND: *In the days before websites or handheld devices, this 1972 schematic representation does a comprehensive job of showing services from four UK domestic airlines (two divisions of BEA plus Northeast and Cambrian). It includes departure and arrival times, flight numbers, days of operations, catering facilities, and even terminals, but is difficult to read.*

LEFT: *A wide-bodied aircraft is able to accommodate at least two aisles with seven people abreast. Known better as a jumbo jet, the first Boeing 747 is seen here at its public unveiling in 1968. It went into service with Pan Am in January 1970.*

OPPOSITE: *As planes got bigger and faster, maps seemed to get less attention. This 1969 Pacific Western map (from its timetable) has destinations and legend quite literally handwritten on.*

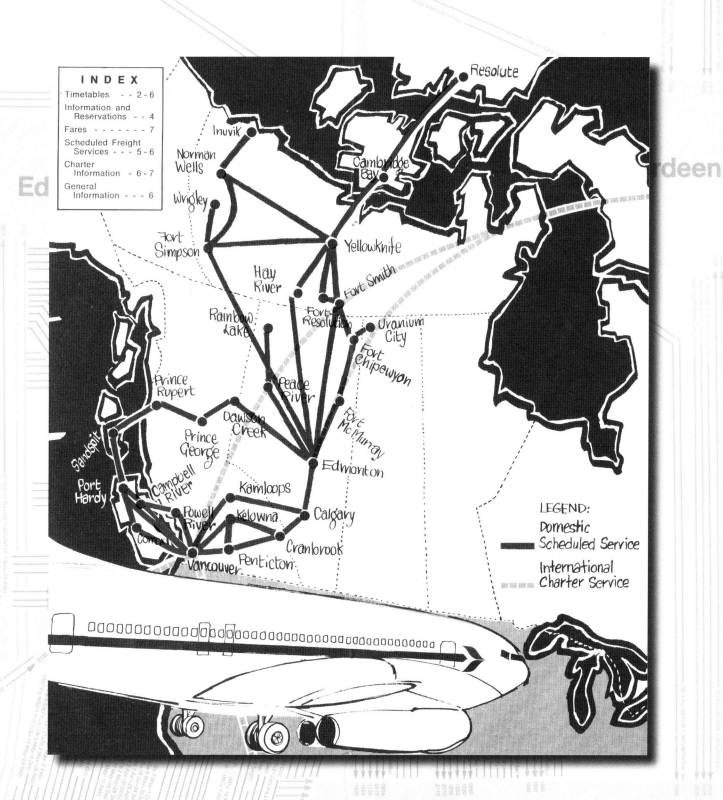

1968–1977

INDEX

Timetables - - 2-6

Information and
  Reservations - - 4

Fares - - - - - - 7

Scheduled Freight
  Services - - - 5-6

Charter
  Information - 6-7

General
  Information - - - 6

Resolute

Inuvik

Norman
Wells

Cambridge
Bay

Wrigley

Fort
Simpson

Yellowknife

Hay
River

Fort Smith

Rainbow
Lake

Fort
Resolution

Uranium
City

Fort
Chipewyon

Prince
Rupert

Peace
River

Dawson
Creek

Fort
McMurray

Prince
George

Sandspit

Edmonton

Port
Hardy

Campbell
River

Kamloops

Powell
River

Kelowna

Calgary

Comox

Cranbrook

Vancouver

Penticton

LEGEND:

Domestic
Scheduled Service

International
Charter Service

Inverness

Wick

Edin

Aberdeen

Manchester

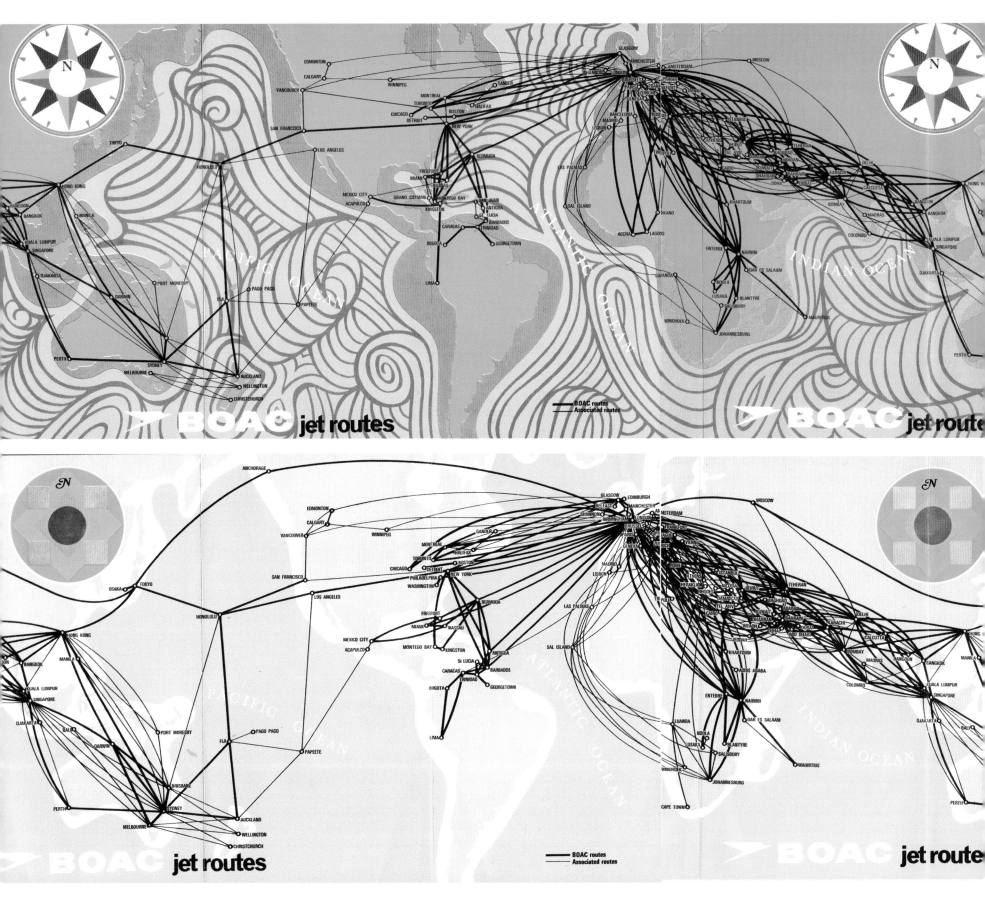

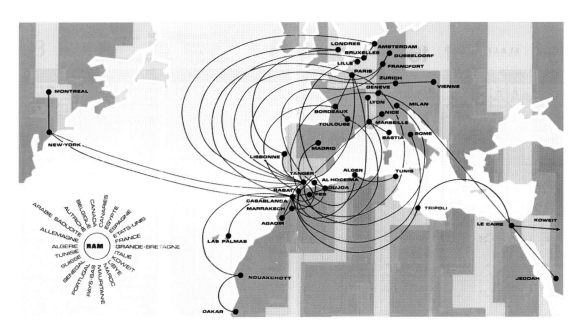

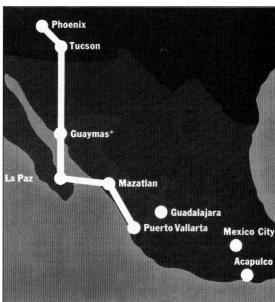

OPPOSITE: *Two examples (neither of which work especially well) of BOAC's international "jet routes." The first, from 1968 (TOP), appears as if the designer may have swallowed some of the psychoactive substances in circulation in the late 1960s. The second, from 1971 (BOTTOM), although blander in hue, is nonetheless beguiling. It is odd that a perfectly legible diagram from a few years earlier (p. 98) was abandoned in favor of these ideas.*
ALL THIS PAGE: *In both architecture and graphic design, an odd development seeped into public buildings and print: the color palette of brown and orange. It was as if some printer or ceramic maker had accidentally over-ordered these hues, giving rise to a spate of publications and architectural finishes so brown and orange that they can be dated to that period by anyone who spots them. The Royal Air Maroc (RAM) map of 1976 (TOP LEFT) and Air West's 1970 one (TOP RIGHT) exemplify these colors—there are many other designs from that era. RAM's map is fascinating for other reasons too: first, the bottom left radial diagram of routes, and second, the trajectory of the lines, deliberately curved to aid legibility, an idea that would be repeated in later years (pp.124–125). Aviaco's 1968 diagram (BOTTOM RIGHT) uses straight lines, and nudges the hue toward the greener side of the spectrum, but its squared-off seas are reminiscent of RAM's checkerboard landmasses (both similar in style to earlier maps, pp. 59–60).*

The battle for the ideological future of jet air travel, mass market versus supersonic, began in 1969. The prototype Boeing 747 made its first test flight in February, and the BAC-Sud Aviation Concorde in March. The increasing popularity of US air travel was beginning to cause congestion of airways and airports, and larger, more efficient aircraft were the obvious solution. As a major customer of Boeing, Pan Am had requested an aircraft with more than twice the capacity of the 707, ordering twenty-five in 1966, and in 1970 it placed into service an aircraft that could carry 276 passengers at a cruise speed of over 550 mph. In contrast, Concorde could carry just 120 passengers, but at an astonishing 1,300 mph, permitting a transatlantic crossing in under four hours.

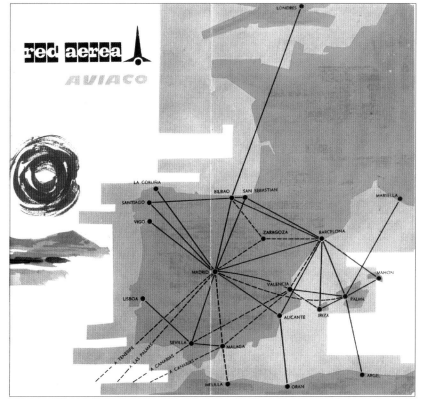

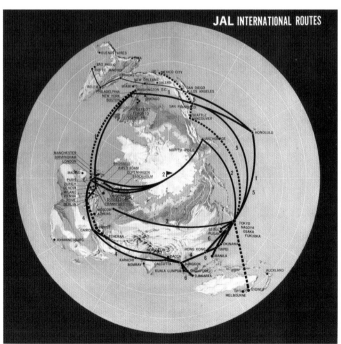

The oil crisis of 1973 settled the issue. Concorde had relatively high fuel consumption, and its sonic booms meant that supersonic speeds would not be permitted over populated areas. In total, international orders for fifty Concorde aircraft were canceled. The ill-fated McDonnell Douglas DC-10 entered service soon after the 747, with similar capacity and speed, and the even larger Lockheed TriStar commenced flights in 1972. Airlines embraced the mass market, and wide-body jets (popularly known as jumbo jets) sold in the hundreds throughout the 1970s.

Wide-bodied jets offered the potential of enormous cost savings: transporting hundreds of passengers in one aircraft instead of two (or even three) saved on crew, fuel, and maintenance. Passengers were not necessarily impressed with airborne mass transit, however, especially as long rows of seats would deprive many of them of window views. Aircraft needed to run to capacity to fully realize the potential economies, but in those uncertain times passenger numbers were slow to rise, especially as the increased fuel prices resulted in higher air fares.

116

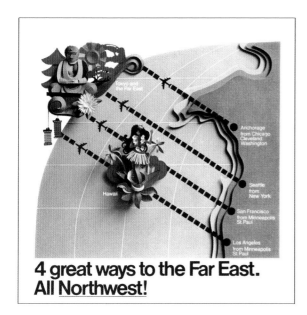

4 great ways to the Far East. All Northwest!

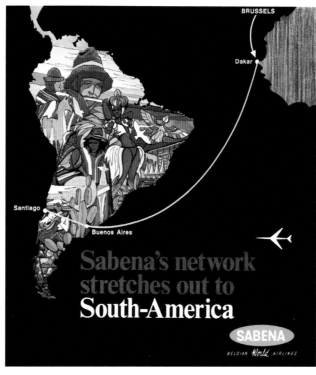

Sabena's network stretches out to South-America

SABENA

BELGIAN *World* AIRLINES

Polar Express

Northwest Express

Aloha Express

Fly the Orient Express Way

Nobody has more routes to the Orient

Nobody flies direct from more U.S. cities (11) and has more 747s to the Orient (21 a week). You can fly Northwest 747s from New York through Seattle, from San Francisco through Honolulu, and from Los Angeles through Honolulu. All flights treat you to magnificent Regal Imperial Service. Fly the Orient Express Way — your choice of 3 routes to 7 Orient cities. For reservations, call your travel agent or Northwest Orient.

Discover the new magnificent

NORTHWEST ORIENT 747

We give you Hong Kong, Singapore, Taipei, and Bangkok.

⊛ Philippine Airlines

TO THE CROSSROADS OF THE ORIENT. MANILA.

ALL OPPOSITE: *The age-old struggle of how to represent the globe on an airline map was ever present in this era. But something incredible happened on December 24, 1968: the Earth was given an early Christmas present, a photo of its entire self from space. Although preempted by the guesses of countless cartographers over centuries (and especially airline route mappers, pp. 66–67, 83), this outstanding image, now known as "Earthrise," had a huge impact, not just on the population as a whole, but on airline route map design in particular. Hence this hastily assembled press ad from the clever design department at Philippine Airlines in 1969 (TOP LEFT). Artistic representations of the planet continue to this day, of course, and the 1974 press ad (TOP RIGHT) from Northwest Orient allowed for a jumbo jet to be shown in all its glory—although for it to have flown this far out would have required a bit more than the "delicious American entrées" on offer. It seems to be heading for the moon. Despite the new realism of "Earthrise," cartographers also continued to wildly distort geography in order to show their routes. This projection (BOTTOM LEFT) was used for several JAL leaflets and timetables in the late 1960s and early 1970s. This one is from a 1968 brochure.*

ALL THIS PAGE: *As competition heated up across the planet, airlines stepped up their publicity campaigns in magazine and newspaper advertisements. The early 1970s saw a plethora of new ideas being attempted. For example, this Northwest Orient ad from 1969 (TOP LEFT) appears to be using folded paper (origami) for promoting its transpacific routes. SABENA chose an oddly black South Atlantic to emphasize its Argentina and Chile route (TOP RIGHT). Northwest Orient went photo-postcard based in 1972 (BOTTOM LEFT), with simplified air routes overlaid, almost matching the famous Vignelli New York City Subway map, also introduced that year. In 1975 (BOTTOM RIGHT), Philippine Airlines went for an open giant clam shell filled with a diagram of routes.*

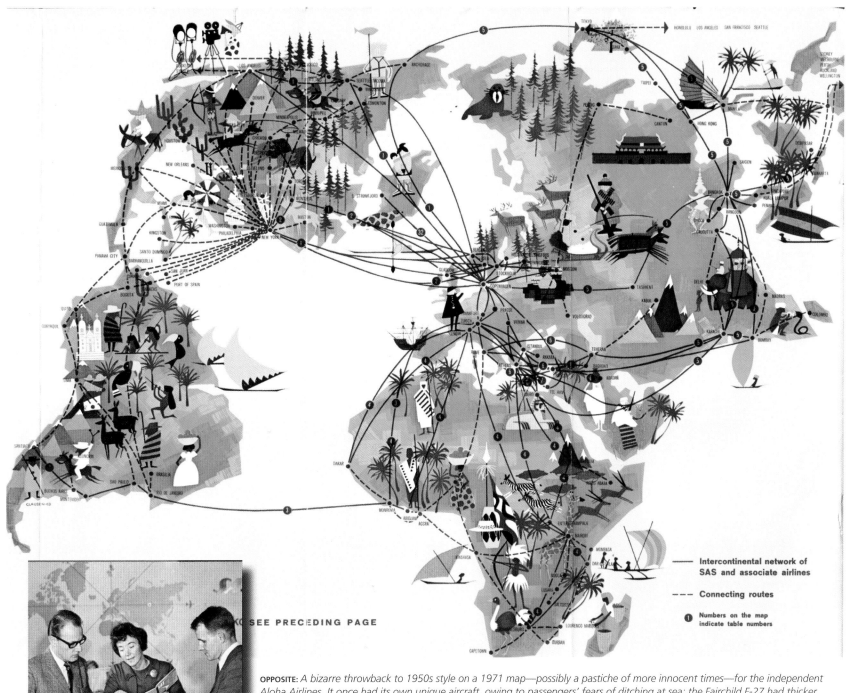

SEE PRECEDING PAGE

OPPOSITE: *A bizarre throwback to 1950s style on a 1971 map—possibly a pastiche of more innocent times—for the independent Aloha Airlines. It once had its own unique aircraft, owing to passengers' fears of ditching at sea: the Fairchild F-27 had thicker underbelly skin and a stronger keel beam than its Fokker cousins. These were replaced by jets in 1967.*

ABOVE AND LEFT: *For a 1968 design, this SAS timetable map (with photo INSET) also has a retro feel and, with the exception of a throwback moment a decade later (pp. 120–121), was probably the last of the major airline maps showing vignettes to this degree. Its quasi-azimuthal equidistant projection (with straightened up Africa) was not uncommon, and an idea SAS often explored.*

Intercontinental network of SAS and associate airlines

--- Connecting routes

Numbers on the map indicate table numbers

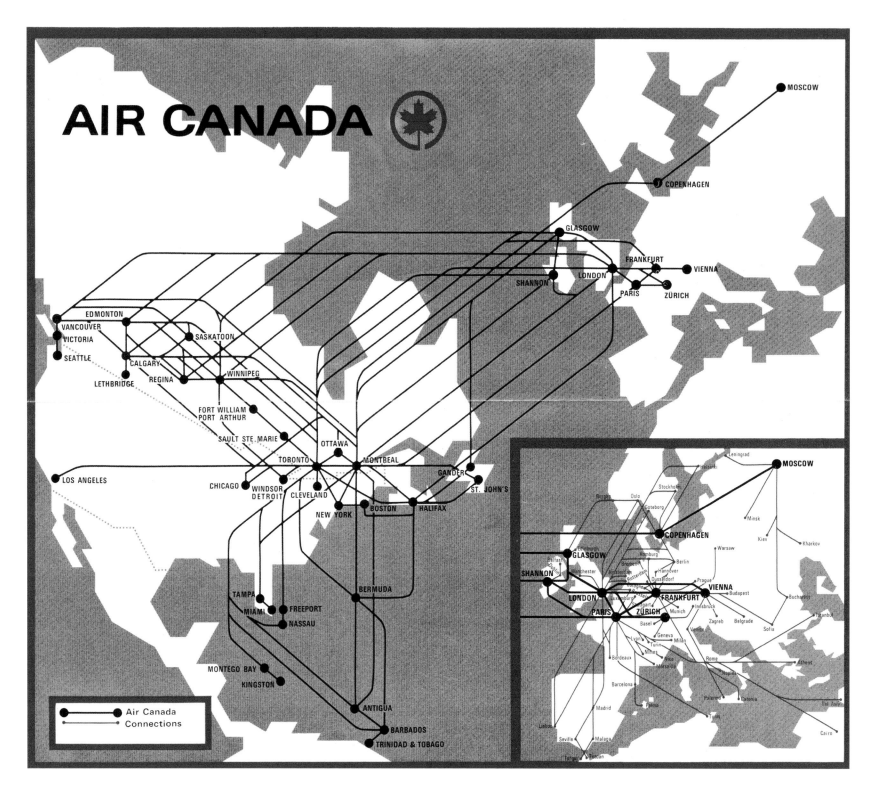

INTERNATIONAL ROUTES

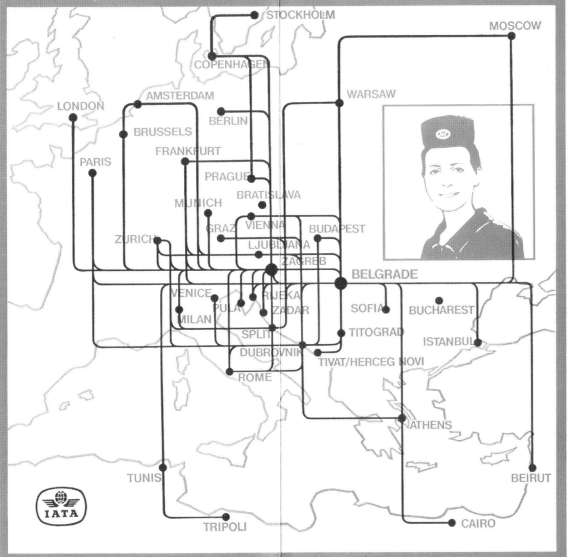

STOCKHOLM
MOSCOW
COPENHAGEN
WARSAW
LONDON
AMSTERDAM
BERLIN
BRUSSELS
FRANKFURT
PARIS
PRAGUE
BRATISLAVA
MUNICH
GRAZ VIENNA BUDAPEST
ZURICH LJUBLJANA
ZAGREB
BELGRADE
VENICE RIJEKA
PULA ZADAR
SOFIA BUCHAREST
MILAN
SPLIT TITOGRAD
ISTANBUL
DUBROVNIK
TIVAT/HERCEG NOVI
ROME
ATHENS
TUNIS
BEIRUT
IATA
TRIPOLI
CAIRO

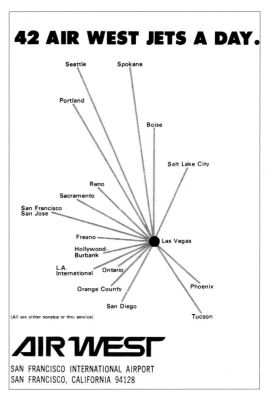

42 AIR WEST JETS A DAY.

Seattle Spokane
Portland
Boise
Salt Lake City
Reno
Sacramento
San Francisco
San Jose
Fresno Las Vegas
Hollywood-Burbank
L.A. International Ontario
Orange County Phoenix
San Diego
Tucson

(All are either nonstop or thru service)

AIR WEST

SAN FRANCISCO INTERNATIONAL AIRPORT
SAN FRANCISCO, CALIFORNIA 94128

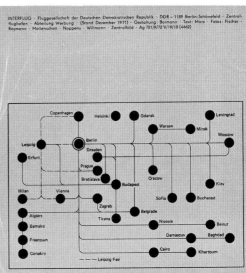

INTERFLUG · Fluggesellschaft der Deutschen Demokratischen Republik · DDR – 1189 Berlin-Schönefeld · Zentral-flughafen · Abteilung Werbung · (Stand Dezember 1971) · Gestaltung: Bormann · Text: Marx · Fotos: Fischer · Reymann · Mollenschott · Noppens · Willmann · Zentralbild · Ag 721/9/72 V/19/18 (4462)

Copenhagen Helsinki Gdansk Leningrad
Warsaw Minsk Moscow
Leipzig Berlin
Erfurt Dresden
Prague Cracow Kiev
Bratislava Budapest
Milan Vienna Sofia Bucharest
Zagreb
Algiers Belgrade Nicosia Beirut
Bamako Tirana
Damascus Baghdad
Freetown
Conakry Cairo Khartoum
Leipzig Fair

ALL ON THIS SPREAD (AND OVERLEAF): *The decade saw a variety of methods to convert topographical reality into abstract diagrams or schematics. After initial interest in the 1930s (pp. 35–37), there was a resurgence in the mid-1960s (p. 98) that has proved enduring. This well-executed 1968 Air Canada diagram (OPPOSITE) led the way for the airline right up to the present day (p. 141). JAT Airways became Air Serbia in 2013, but its 1973 design and inset stewardess (ABOVE) graced timetables for some years. East German airline Interflug (BOTTOM RIGHT) used similar forms—just with larger dots and no coastlines—on this diagram from the same year. Air West had already dispensed with surface features on this 1970 radial schematic (TOP RIGHT).*

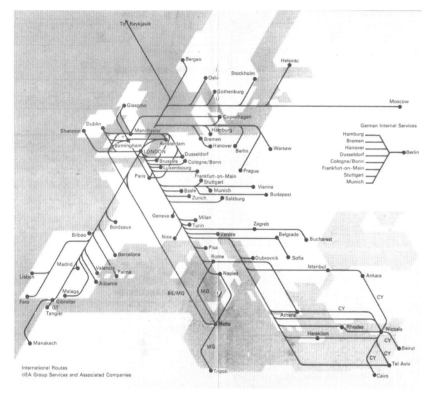

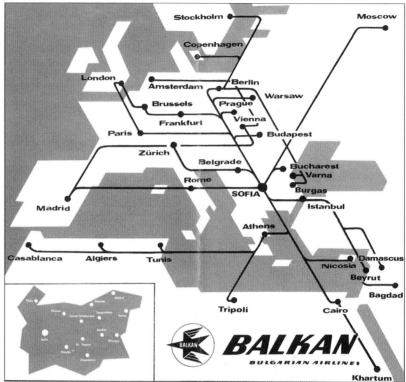

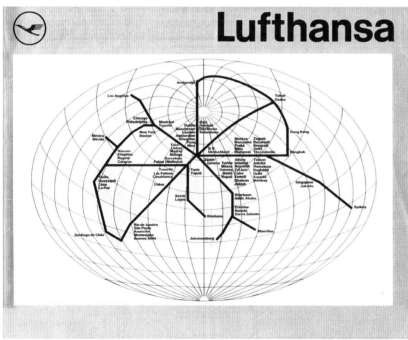

AEROFLOT *Soviet airlines*
spreads its wings both East and West

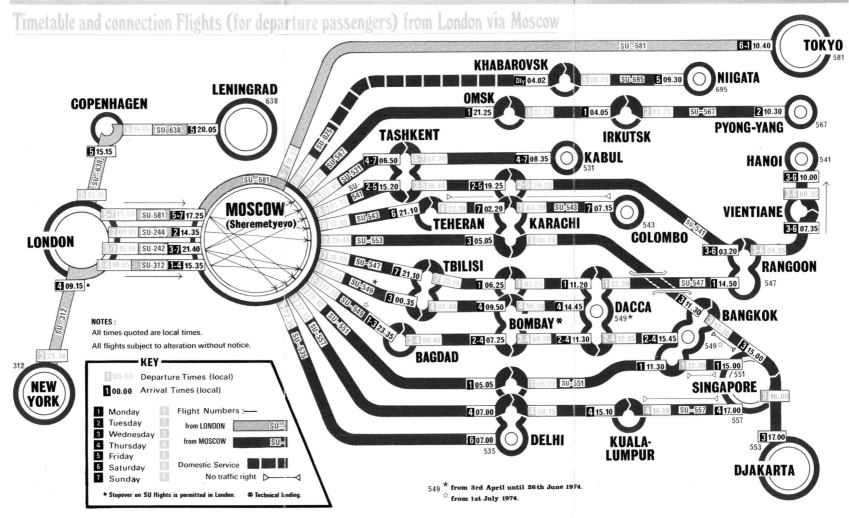

Timetable and connection Flights (for departure passengers) from London via Moscow

NOTES :
All times quoted are local times.
All flights subject to alteration without notice.

KEY

00.00 Departure Times (local)
00.00 Arrival Times (local)

Flight Numbers :—

1 Monday
2 Tuesday
3 Wednesday
4 Thursday
5 Friday
6 Saturday
7 Sunday

from LONDON   SU–
from MOSCOW   SU–

Domestic Service
No traffic right ▷———◁

* Stopover on SU flights is permitted in London.   ✿ Technical landing.

549 ★ from 3rd April until 26th June 1974.
    ☆ from 1st July 1974.

ALL ON THIS SPREAD (AND PREVIOUS): *No apology is made for the focus on schematics during this era; they were without doubt a dominant cartographic form used in the 1970s. By the time British European Airways published this stylish 1971 diagram (*OPPOSITE TOP LEFT*), discussions were already taking place about possible mergers. Its internal West German routes, started in 1968, are also shown, along with a simplifying trajectory of a single route line (London to Naples) with branches off. A similar idea was employed on a 1974 Balkan diagram (*OPPOSITE TOP RIGHT*) for Berlin to Cairo. Lufthansa's 1975 world routes takes a different schematic approach (*OPPOSITE BOTTOM LEFT*). While the 1975 Garuda (*OPPOSITE BOTTOM RIGHT*) is curvaceous, it is also a diagram. In 1974, Aeroflot (*ABOVE*) included operational details (departure and arrival times, flight numbers, days of the week, etc.) to a clearer degree than had been achieved earlier with similar schematics (pp. 45, 78–79, 102–103). Note that although the map is presented in the English language, it has some interesting spelling (Vientiane, Dacca, Djakarta).*

Last year Peking & Tokyo.

This year, New York.

(From May 29)

New Yorkers will see the 'Homa', ancient Persia's legendary bird, <u>five times weekly</u> on the tailplanes of Iran Air's all-Boeing fleet.

Our 'Homa' network now embraces places as far apart as New York, London, Tehran, Peking and Tokyo. Ultra modern Boeing comfort and sky high standards of passenger care are helping to make Iran Air one of the fastest growing names in the airline world.

IRAN AIR

New York, London, Paris, Frankfurt, Vienna, Geneva, Zurich, Rome, Moscow, Athens, Istanbul, Tehran Abadan, Baghdad, Kuwait, Bahrain, Abu Dhabi, Dhahran, Dubai, Doha, Muscat, Kabul, Karachi, Bombay Peking, Tokyo. Also sales offices in Milan, Hamburg, Los Angeles and Houston.

هواپیمائی ملی ایران . هما

Britain's right hand men choose the right airline

WINTER TIMETABLE

AIR ANGLIA

Effective 27th October 1974
Until 31st March 1975   Subject to alteration without notice.

ALL OPPOSITE: *Aside from the orange and brown hues of the early 1970s, neon yellow and sky blue also often went together. A 1973 Rocky Mountain Airways timetable cover (OPPOSITE TOP LEFT), South American operator LACSA's 1971 press ad (OPPOSITE RIGHT), and an Air Canada 1972 ad (OPPOSITE BOTTOM LEFT) all used the colors.*

LEFT: *Iran Air's 1974 press ad for network expansion, sadly short-lived.*
RIGHT: *Although it never operated any wide-bodied jets, little Air Anglia did embrace the mid-1970s color palette and cartoon style.*

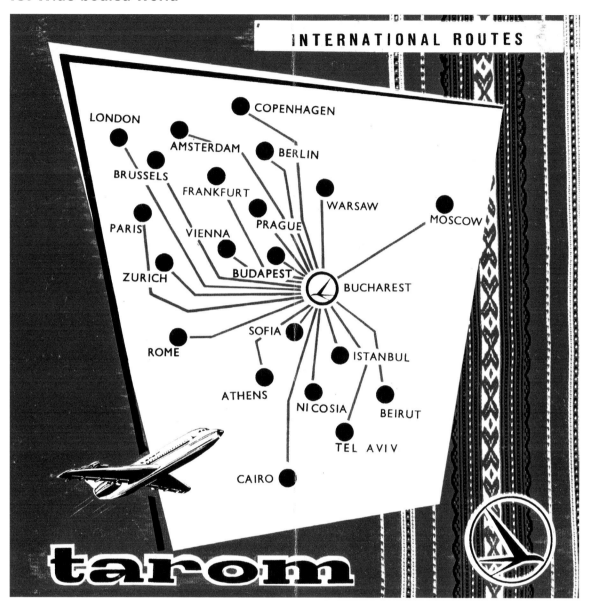

LEFT: *Although Tarom can trace its routes back to the 1920s, it began operating under the name Transporturi Aeriene Române in 1954 as the state-owned airline of Romania. It had started using the Soviet designed and built long-range, narrow-bodied jet Ilyushin IL-62 in 1973, around the time that this diagram was produced. The jet shown on this map is the British-built BAC-111, however, which Tarom also purchased. By the late 1970s, Tarom had the largest Soviet-bloc fleet outside of Aeroflot.*

ABOVE: *Aeroflot itself was huge, claiming to be the largest airline in the world when this cover for its 1975 map was released.*

OPPOSITE: *Two Aeroflot maps from the same year, 1975, seem to serve different markets. The first—for Western tourists, perhaps—evoking a golden age of airline route maps covered in happy, smiling vignettes (**TOP**), the other—far more austere—a return to a more schematic form and, interestingly, with no place names (**BOTTOM**). Although barely visible on these items (perhaps played down for the international market), the Aeroflot logo contains a nested winged flying hammer and sickle—and it is still part of the company's insignia today.*

From a marketing point of view, bigger is less exciting than faster, and it is noticeable that the glamor and excitement of past decades has faded away. Even the colorful vibrancy of the 1960s proved short-lived (p. 104), replaced by a blander outlook (pp. 104–105). The commercial artistry that graced magazine advertisements for three decades had also reached the end of the line (p. 106), superceded by photography (pp. 106–107, 115) joining the comical drawings (pp. 114–115) first seen in the previous chapter (p. 101). Four pictorial maps run the full range, from gloriously anachronistic (p. 108) to attractively informal (p. 109) to bizarrely cartoonish (pp. 115, 117). The abstract diagram is ascendant in this era, worldwide, with a variety of schematization techniques applied to simplify the complicated networks (pp. 110–113), especially from the countries behind the Iron Curtain, and with more than a nod to the London Underground diagram as one of the early pioneers of turning complex routes into simple straight lines (p. 113).

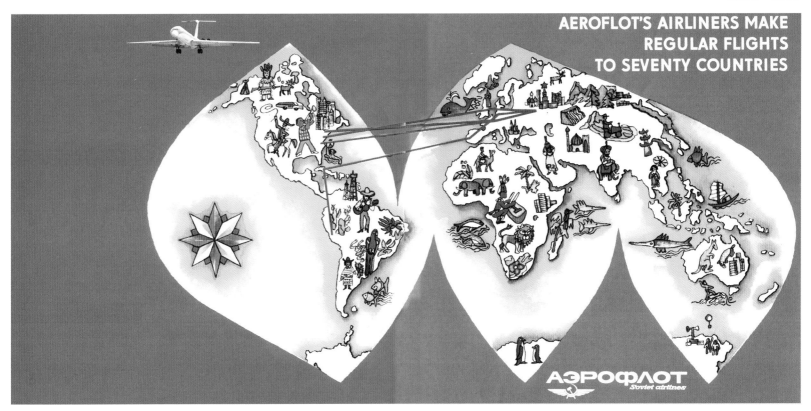

AEROFLOT'S AIRLINERS MAKE
REGULAR FLIGHTS
TO SEVENTY COUNTRIES

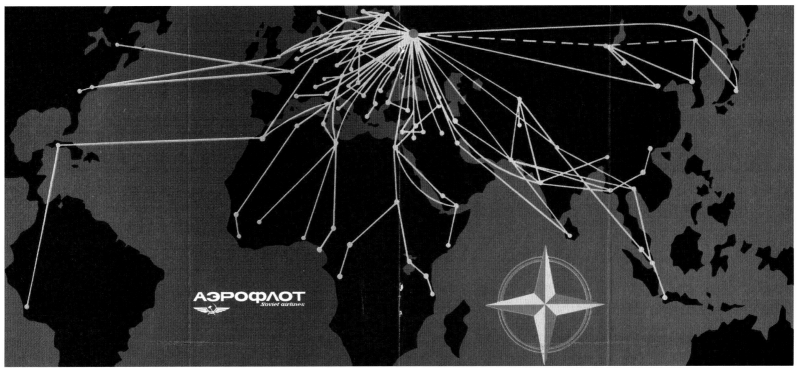

Terminal 6: Open skies for all

BACKGROUND: *A 1980 map of British Airways' European routes promoting where its new TriStars were flown.*
LEFT: *Another low-cost airline to rapidly grow was EasyJet, founded in 1995. The first call center number painted on plane sides remained on the livery for several years.*
OPPOSITE: *A 1999 map from Irish-based Ryanair, now one of Europe's most successful airlines. It began in 1985 but was a major beneficiary of 1997 deregulation laws.*

Air deregulation commenced in the United States in 1978 and rapidly spread worldwide. The giant flag carriers could no longer rely on government regulation to ensure their profits. Maps continued to be important publicity tools, and newer operators used them proudly to advertise their competing services. The need to attract business led to a spike in creativity, compared with maps in the previous chapter, perhaps assisted by the introduction of computer graphics that democratized graphic design. Sadly, many famous names were unable to adapt to the new business environment, and their maps appeared for the last time during this era.

1978–1999

1979 ROUTE MAP

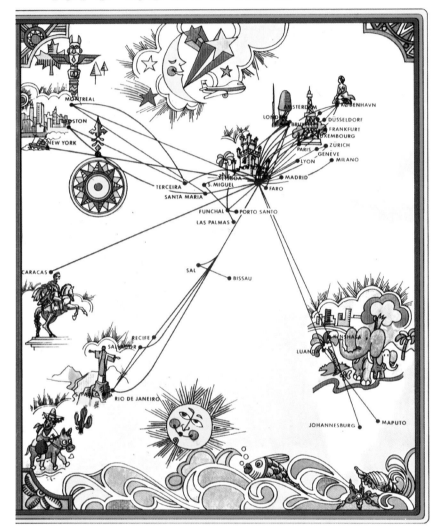

Go Family Island hopping with a member of the family.

Bahamasair wants you to come meet the entire family—all twenty of our beautiful Family Islands. There's Eleuthera, The Abacos, The Exumas and seventeen other exotic destinations waiting for you. And we're the only airline that flies to all of them.

Of course we also fly our 737's to Nassau and Freeport from Miami, Ft. Lauderdale and Atlanta. So whether it's snorkeling and swimming or golfing and gambling, Bahamasair is your best way to go.

For reservations call: Continental U.S.A. 800-327-8080; Florida (excluding Miami) 800-432-1375; Miami, 442-8585. Or see your travel agent. And remember, when it comes to the Family Islands, we don't just hop there. We live there.

Bahamasair
We don't just fly there. We live there.

ALL ON THIS SPREAD: *Although vignettes of local landmarks or people had been a staple output of airline cartographers since the 1920s (p.13), they had somewhat fallen out of fashion by the 1980s. This selection demonstrates that these rare frivolities, where they were still to be found, would often verge on the infantile. Take the 1979 poster from Portugal's airline TAP (ABOVE LEFT). It feels like it may be more at home in nursery. A 1982 press ad from Bahamasair could just as easily be an illustration for a children's book, with its wacky smiling jet bouncing between the islands (ABOVE RIGHT). Such styles were not uncommon, and the practice resurfaced in the 1990s: the 1996 map for Georgia-based budget airline ValuJet (OPPOSITE LEFT) stretches the elastic the farthest yet, with a professional cartoonist employed to draw the entire design. The fun and games are still present in a similar vein with a 1997 ad by Canadian Pacific based on a pinball machine (OPPOSITE RIGHT). TAP remains a respected airline, and experimented with new designs, but it never returned to such whimsical levels. Neither did Bahamasair, which is still operating. Canadian Pacific—once the country's second largest airline—was absorbed by Air Canada in 2001, but it too steered clear of fanciful marketing after this. The same cannot be said for ValuJet, which insisted on keeping the cartoon mascot ("the Critter") in its logo, even after a crash that killed 110 people. The map shown here was issued soon after the fatal accident, which enraged those who lost family in the tragedy. ValuJet ceased trading in 1997.*

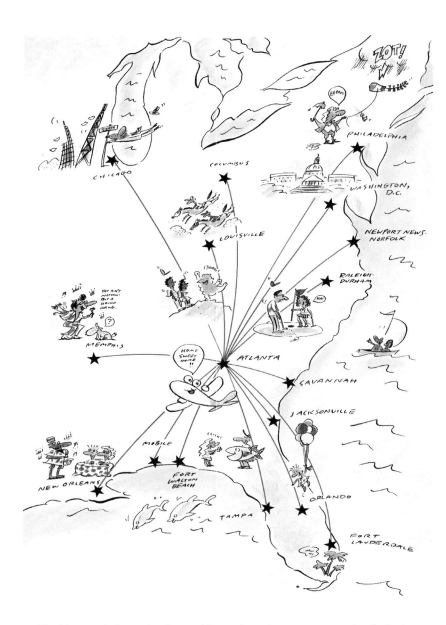

The biggest shake-up in air travel in modern times was not technological, it was political. Until 1978, flag-carrying international airlines were generally treated as representatives of their respective countries, to be protected no matter what, and often under state control. In earlier days of flight, it is understandable that there might have been a strategic need to protect the countries' interests in a world that had twice been at war within the space of twenty-five years. The long-term consequence of this was that on many routes fares were regulated and new entrants were restricted, shielding the established operators from competition and promoting stability.

By the 1970s, air services effectively had a monopoly on long-distance travel, and so the nature of its business inevitably changed. There were no opportunities left to expand by gobbling up train and ocean liner passengers, and so the only ways in which an airline could increase its numbers would be by encouraging more people to travel, others to travel more often, or by raiding passengers from other operators. With regulated fares, the scope for raiding was limited. Hence, the established airlines continued to maintain standards for food, seating, and service; cutting costs might have made their flights more profitable, but might also have driven passengers away to other similarly priced companies that were

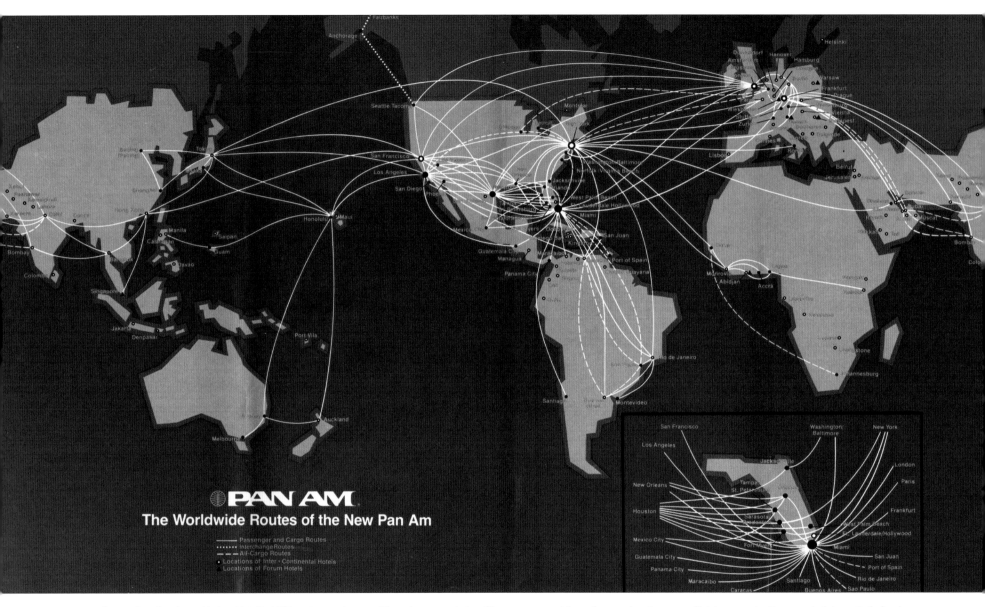

ABOVE: *From a Pan Am timetable booklet, its 1981 system route map. The colors are calming. The common problem of maps showing many flights concentrating on a hub is partially fixed by the inset lower right, which adds clarity to the flights to and from Miami. Pan Am ceased trading in 1991.*

determined to maintain standards. With the stagnation of growth in passenger numbers in the 1970s, however, the old protectionist ways appeared to be encouraging semi-monopolistic practices that kept fares artificially high. Charter flight operators had shown that it was possible to circumvent regulated fares. One in particular, Laker Airways, had acquired considerable expertise in cutting operational costs across the board, showing that carefully managed aircraft could offer cheap places to prebooked blocks of passengers and still be profitable. Indeed, the aircraft were managed so carefully that the company acquired a reputation for exceptional maintenance standards. Laker Airways had fought for many years to offer low-priced scheduled transatlantic flights,

129

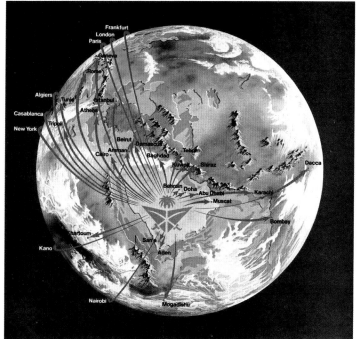

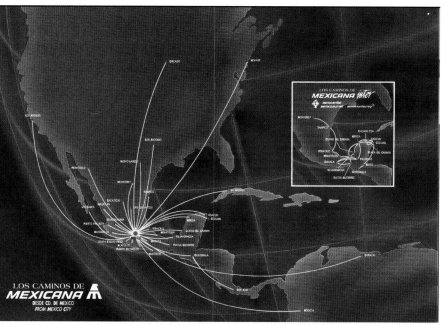

ALL ON THIS PAGE: *A variety of globes in the era of deregulation, some dramatically enhanced with the latest computer graphics, recalling airbrushing from past eras.*
TOP LEFT: *Saudi Arabian Airlines was established in 1945 and changed its name to Saudia in 1972. This press ad dates from 1980, when a number of new long-haul routes had commenced.*
TOP RIGHT: *Avianca had merged with two smaller operators in the year before this stylish 1995 timetable cover was published. By placing Colombia at the center of the globe, routes naturally radiate from Bogotá. Avianca celebrates its centenary in 2020.*
BOTTOM RIGHT: *Another long-established airline in the region was Compañía Mexicana de Aviación, S.A. de C.V. (aka Mexicana), which began operating in 1921. This 1993 route map (with similar hues to Avianca, above it) also uses a globe base but is even more focused on the home country, Mexico. It ceased operations in 2010.*

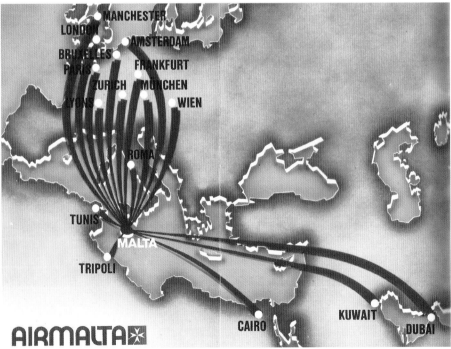

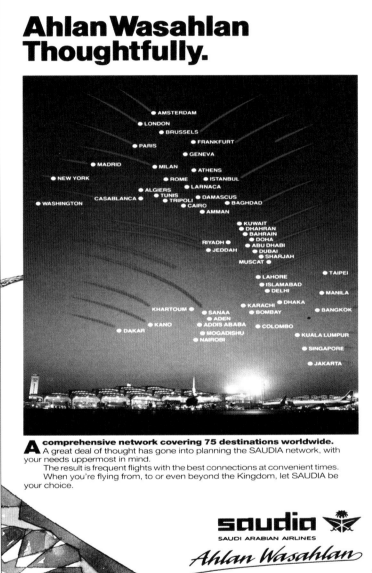

TOP LEFT: *Republic Airlines began operating just two years before this clever map was issued in 1981. The neon route trails appear computer generated, though this is unlikely given the date, but the looping effect certainly stayed around.*
LEFT: *Air Malta was founded in its present form in 1973. Its early 1980s route map seems to be positively bursting out of the island—or targeting it!*
TOP RIGHT: *Saudia tried yet another approach on this 1990 press ad, with what feels like shooting stars or flares heading to the desert lands.*

I N S I D E

Details on Pacific Western Holidays' Trans-Canada charter flights for summer 1983. Departures from Vancouver, Calgary, Edmonton, Winnipeg and Toronto. Pacific Western Holidays feature Boeing 737 and the advanced Boeing 767 aircraft.

Pacific Western

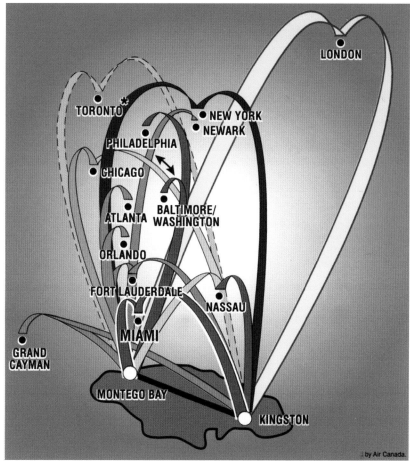

by Air Canada.

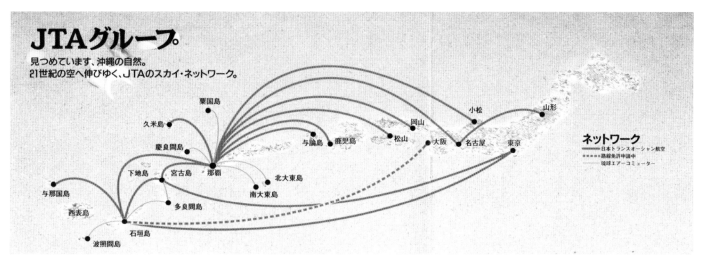

JTAグループ。

見つめています、沖縄の自然。
21世紀の空へ伸びゆく、JTAのスカイ・ネットワーク。

ネットワーク
—— 日本トランスオーシャン航空
••••• 路線免許申請中
—— 琉球エアーコミューター

TOP LEFT: *In a design somewhat echoing the Republic ad (***OPPOSITE***), Pacific Western uses loops to show its 1983 routes on a timetable booklet.*
TOP RIGHT: *More loops, which this time look akin to colored ribbons, on this joyful 1996 map by Air Jamaica.*
LEFT: *JTA also employs loops to bounce between the many Japanese islands in this 1994 magazine ad.*

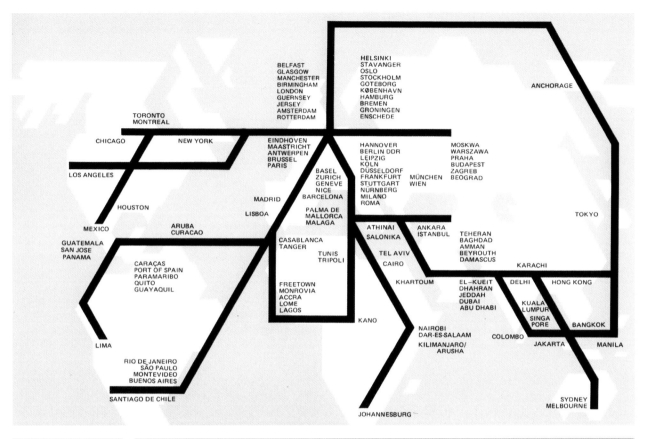

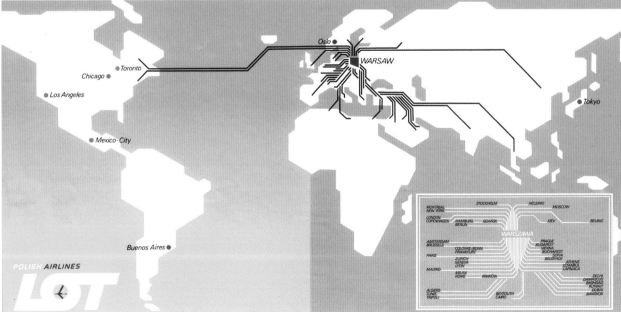

ALL ON THIS SPREAD: *In a decade devoted to diagrams, this spread sums up the various attempts to simplify complex services, all with a blue theme. KLM's 1980 world map gives only generic corridors, with airports listed alongside, so the passenger would need to consult other material to work out the details of a journey (***TOP LEFT***). Polish airline LOT employed a similar method, with even fewer destinations listed on the main area of this 1987 corporate brochure (***BOTTOM LEFT***). There was, however, an inset with a more detailed breakdown of European routes. Amsterdam's Schiphol Airport made this pretty diagram for connections to the UK in 1980 (***OPPOSITE TOP LEFT***). A couple of years later, KLM had added three Irish services, and it produced this diagram of routes from its Amsterdam hub in 1983 (***OPPOSITE TOP RIGHT***). Although the CSA (Czech Airlines) diagram was produced earlier, this late 1970s version sits well among its younger company. (***OPPOSITE BOTTOM LEFT***). It also shows another style of diagram, the spoke or radial concept used by some others (p. 111). A different approach was utilized by Aviaco in 1986 (***OPPOSITE BOTTOM RIGHT***). Here, all diagonals have been banished in favor of just horizontals and verticals. The landmass background is also of note: petite silhouetted jets.*

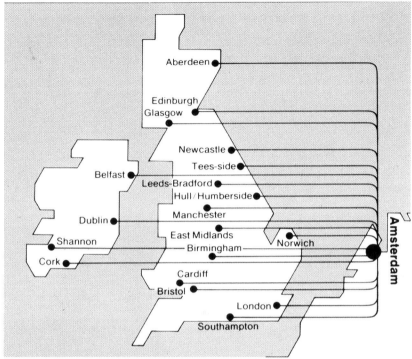

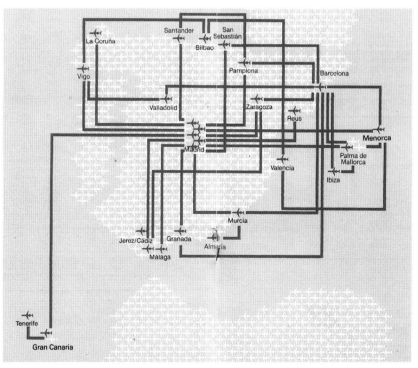

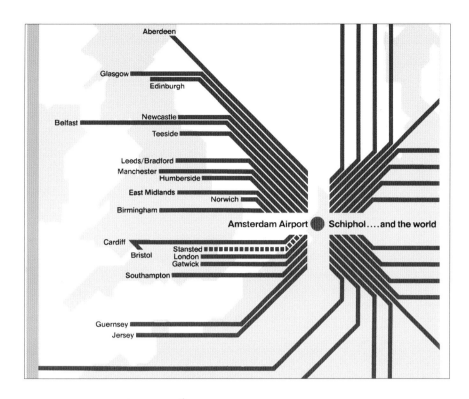

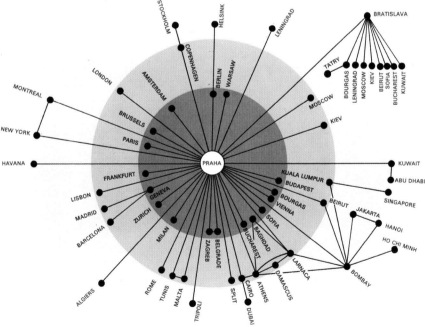

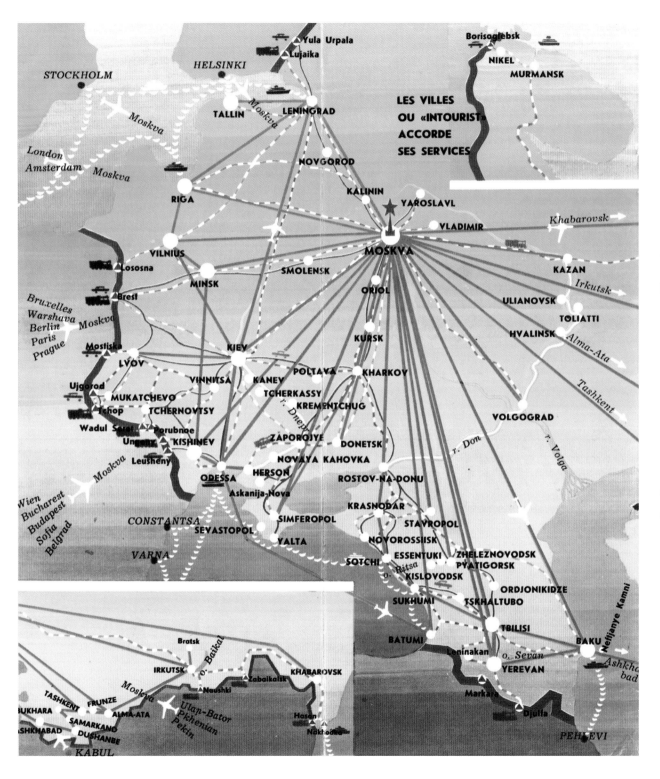

LES VILLES
OU «INTOURIST»
ACCORDE
SES SERVICES

AirUK
TIMETABLE
27 March to 29 October 1988 Issue One

Domestic & International Scheduled Services

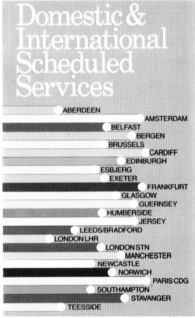

LEFT: *In preparation for Moscow hosting the 1980 Summer Olympic Games, the official agency for attracting visitors, Intourist, raised its profile. This colorful map was produced in the following year, 1981, for the French-speaking market, and it shows just how much of the country was being opened to tourists, even before the Glasnost era.*

ABOVE: *AirUK was a small independent British airline that began operations in 1980. This colorful and cute cover is not so much a diagram as an alphabetical list of routes.*

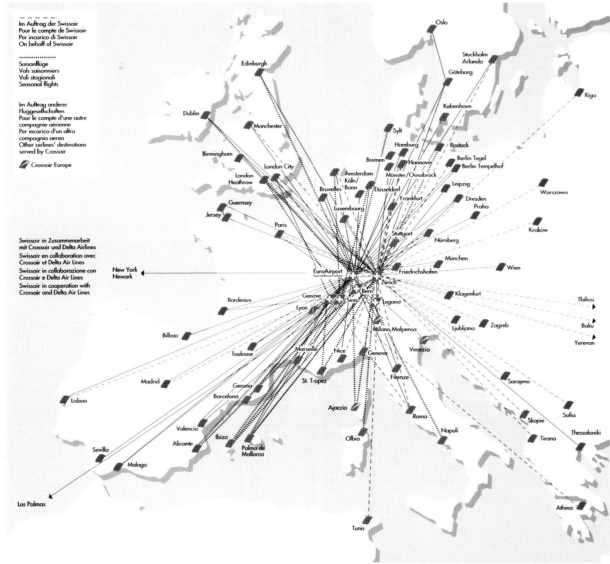

ABOVE: *This 1991 leaflet from TAP shows similarities with the AirUK timetable cover (OPPOSITE). The list of destinations is placed alphabetically. But because the red squares are not linked to the lists, it is of limited use unless the reader is a geography expert.*

RIGHT: *The 1999 Crossair diagram also fails the legibility test. To follow routes out of its Swiss hubs (especially EuroAirport at Basel) is tricky. The airline began in 1978 and was renamed Swiss International Airlines after taking over Swissair in 2002.*

and it was finally granted a license to offer a daily flight from London Gatwick to JFK. Operations began under SkyTrain branding in September 1977. The eagerness of new operators to offer reasonable no-frills services at considerably lower fares, under $250 for a SkyTrain transatlantic trip versus over $600 for a standard economy class fare, must have been the ammunition that laissez-faire economists, inspired by Milton Friedman, were looking for.

Soon after, the US Airline Deregulation Act of 1978 removed regulatory barriers to new operators on new routes, ending minimum prices for fares. Across the world, deregulation of domestic markets in many countries followed, alongside open-skies agreements between countries. The repercussions

worldwide have been massive. The liberalization of the airline marker spurred many new operators. Maps from Ryanair, ValuJet, Republic Airlines, and Crossair appeared for the first time in the era covered by this chapter. ValuJet did not survive this era, and long-established Braniff was, unfortunately, unable to adapt. Other names, such as Pacific Western, vanished in a maelstrom of mergers and aquisitions.

The most famous fall from grace was Pan Am, which paid a high price to acquire National Airlines in 1980, and then went into a tailspin of asset sales in an attempt to stay viable before succumbing to bankruptcy in 1991, despite maintaining passenger numbers. Against the backdrop of corporate turmoil,

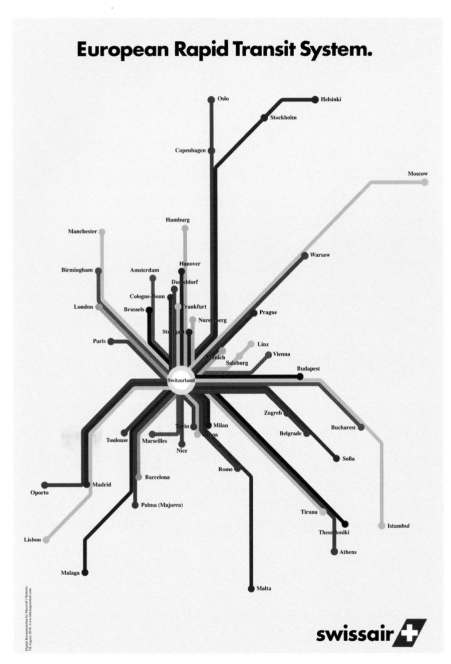

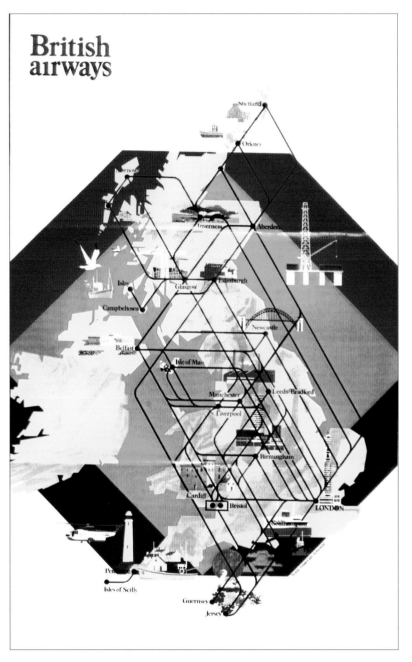

ABOVE: *The 1987 Swissair route diagram advertisement has achieved near-legendary status among cartophiles. It shows only the company's European destinations and is similar in style to the famous New York City Subway map of 1972 by Italian designer Massimo Vignelli (1931–2014). Vignelli had no known connection with Swissair, but his favored typeface, Helvetica, links him to Swiss design. This version is a digitization by Maxwell Roberts.*

ABOVE: *British Airways was created in 1974 from a merger of BEA, BOAC, and two regionals. The airline launched an advertising campaign called "Fly The Flag" in the mid-1970s, the logo of which consisted of a blue and red lozenge. This device forms the backdrop for a colorful British Isles linked by horizontals and 60-degree diagonals on this (late 1970s) poster, which may not have been released.*

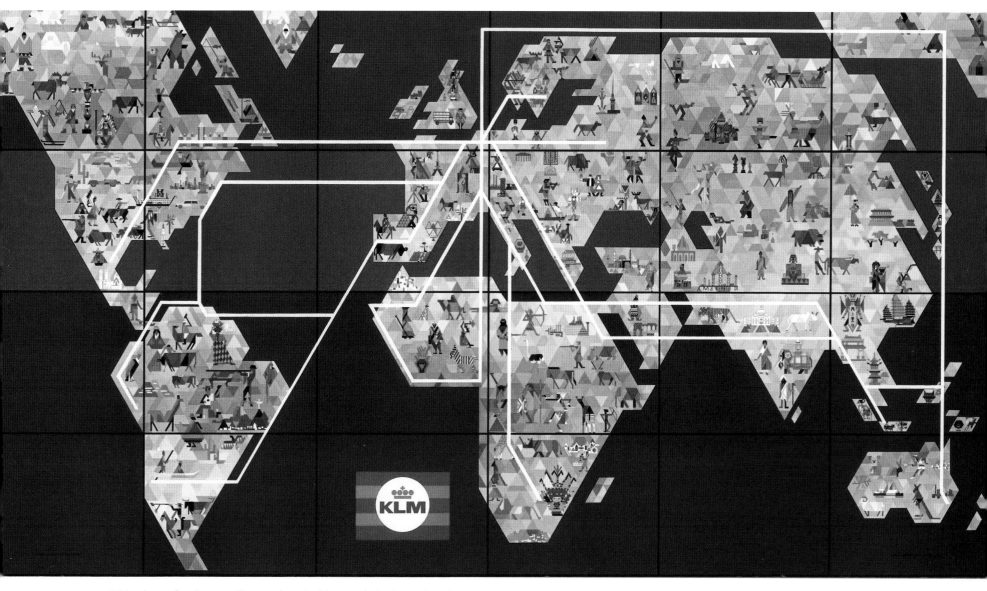

ABOVE: *With echoes of earlier KLM diagrams (p.126), this 1985 design by Studio Osborne & De Karte, based on 60-degree diagonals and equilateral triangles, is a masterpiece.*

there have been fare price winners and losers, and for people prepared to hunt around and be creative, there were indeed bargain flights to be had.

With the new urgency to attract business, the maps in this chapter seem to have more enticing vibrancy to them than those in the previous one. Part of the contribution to this is the development of computer graphics, with capable desktop systems becoming widely available during the 1990s. In conjunction with this, the price of color printing fell considerably, giving a much needed boost to creativity. The lift that color gives, for example, to diagrams has resulted in outstanding works that can withstand comparison with any other maps in the book (pp. 130–132), especially compared with the severe two-color diagrams from the previous era (pp. 110–112). Even the less colorful diagrams seem more accessible than previously (pp. 126–127), and the use of arced trajectories almost recalls the excitement of earlier decades (pp. 124–125), as does the use of globes, especially when enhanced by computer graphics (p. 123).

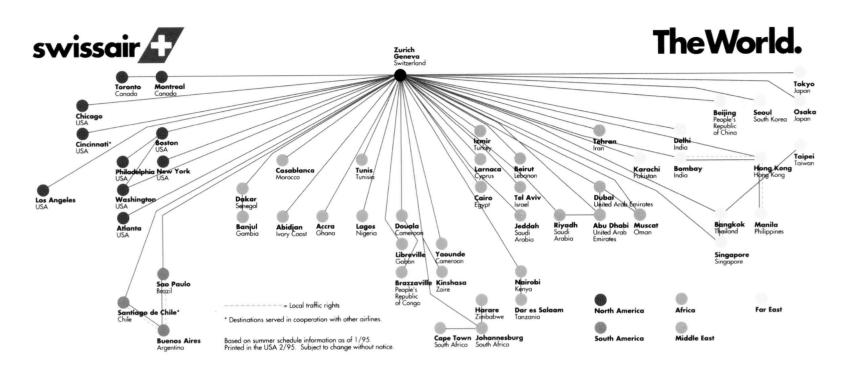

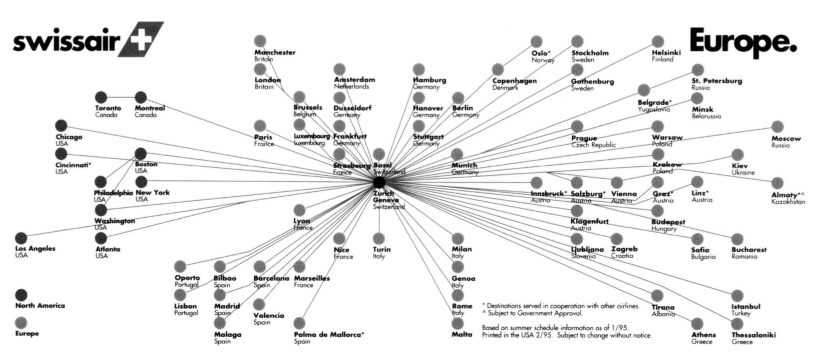

OPPOSITE: *Before the company's demise, Swissair was famous for its publicity material. These two 1995 diagrams, with routes radiating from Zurich, were printed on a glossy plastic card, one each side. They were inserted into seat-backs and were a common sight on its routes in the late 1990s (before the age of built-in TV screens).*

RIGHT: *For sheer simplicity and beauty that sums up an airline's sunny destinations, this 1990 advertising campaign from Iberia is hard to beat. This version is from a German-language press ad. Sonnensystem means "solar system."*

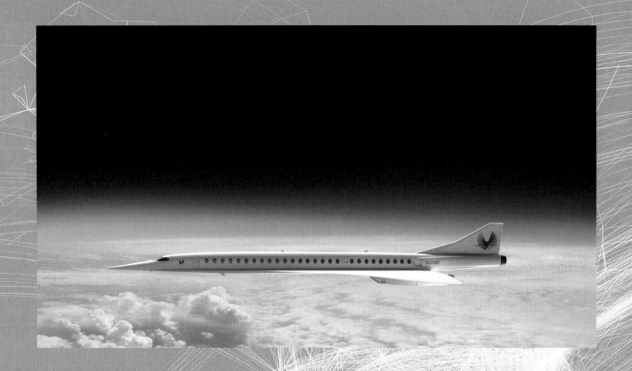

Terminal 7: Low cost to low orbit

Events set in motion in the previous chapter sowed the seeds for the rise of the low-cost airline. Now, for most people, ticket price determines everything, and the cheapest flight can be booked through sites that aggregate fares from hundreds of airlines. Maps seem strangely irrelevant. They are still ubiquitous, and there are still some lovely examples in timetables and publicity materials, but inspirational work is not easy to find. The message seems to be that, in its defeat of all other forms of long-distance transport, air is now the ordinary way to travel, and airline maps no longer inspire the same level of creative energy. Times change, however, and who knows what future developments might arise and rekindle the excitement of flight, and maps that show the way.

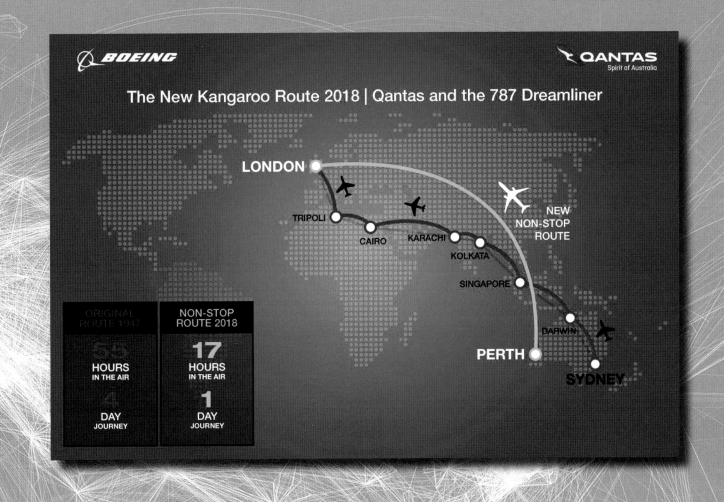

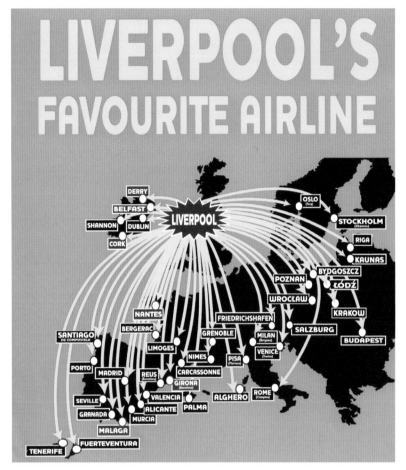

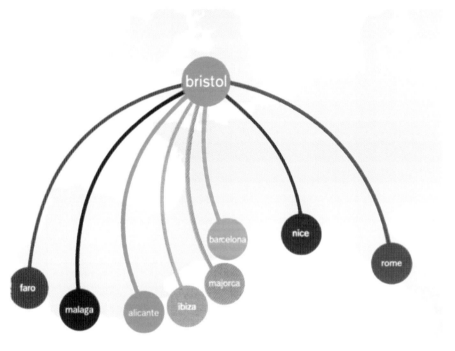

ALL ON THIS SPREAD: *Routes bursting from the center (the ripple effect) is not new in airline mapping, but these twenty-first-century examples show how popular the technique remains. A Ryanair map from the early 2000s (TOP LEFT), Go Airlines from 2001 (TOP RIGHT), Air Berlin from 2000 (BOTTOM LEFT), and FinnAir from 2004 (BOTTOM RIGHT), which has the added bonus of a cloud effect.*

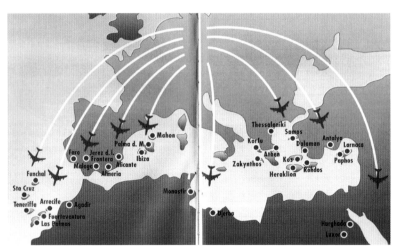

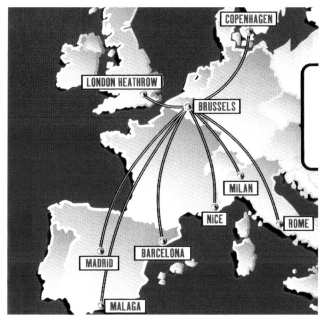

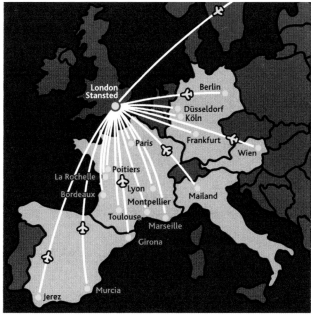

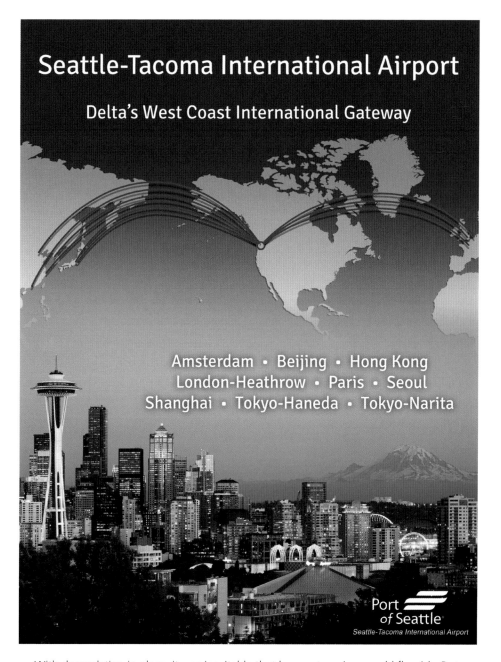

RIGHT: *Few airports, advertising their destinations, have achieved this to such a classy extent as Seattle-Tacoma International, in a 2014 ad designed by Ian Sharma.*
TOP: *The Virgin Express airline was based out of Brussels Airport (1996–2006). This map is from its 2001 timetable booklet. Virgin Express was merged with Brussels Airlines in 2007.*
ABOVE: *A 2000 flier by low-cost airline Buzz (ceased trading in 2004).*

With deregulation in place, it was inevitable that low-cost carriers would flourish. But "low-cost" here refers to an airline streamlining all its operations, saving money at every opportunity, rather than always low cost to passengers. Now that everything is a paid-for extra, a pleasant trip on an aircraft is no longer standard. Despite this, the availability of cheap fares online has proved irresistible, and has transformed air travel, perhaps forever.

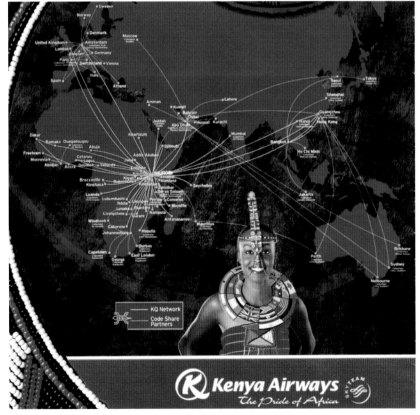

AFRIQIYAH AIRWAYS

VOUS OUVRE LES PORTES DE L'AFRIQUE

ALL ON THIS PAGE: *Sometimes overlooked, Africa was one of the earliest international destinations at the dawn of civil aviation. Founded in 2001, the state-owned Libyan Afriqiyah Airways produced this map in 2005 (**ABOVE**). Geographical accuracy rules here, except Great Britain seems to have vanished, and London shifted to southern Ireland. A 2015 ad (**TOP LEFT**) for Air France and KLM's partnership to better serve "40 countries and 50 cities" across the African continent is symbolic of how such a simple idea of birds in flight can be manipulated digitally to evoke a map. Also from 2015, Kenya Airways' route map (**BOTTOM LEFT**) shows a growing international network, although some lines include code share partners.*

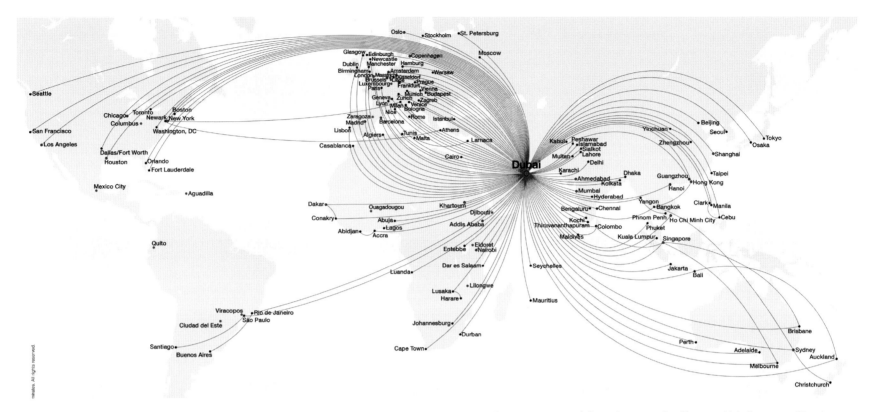

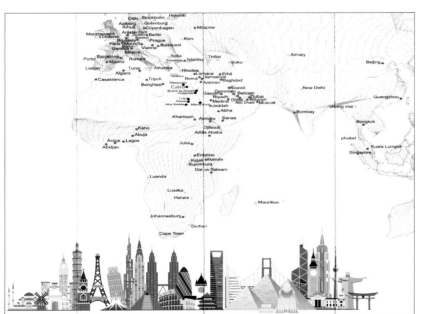

ABOVE: *Emirates' 2018 network is a prime example of how multiple lines can still be shown, provided they all emanate from just one place—and are executed as neatly as this.*
LEFT: *An alternate idea, shown on this EgyptAir map of 2016, is to dispense with the lines for the plane routes and use them in other ways—here, as oddly bizarre contours. The world landmarks are a nice touch, although order and scale are slightly random.*

Yet more famous old names ceased trading since 2000, including Swissair, SABENA, and TWA. Mergers and code sharing, creating massive global alliances, have helped the rest survive.

What of the future? There are extravagant claims of flying to other worlds, but the reality is more prosaic. The Boeing 787 carries around 170 more people than the original 747, at almost exactly the same speeds, although its improved fuel efficiency and range shrinks the globe that little bit more. Eight hundred and fifty people can be squeezed into the slightly slower Airbus A380. Boeing's planned 797 is midsize, with seats for 228/267, carrying more fares than a narrow-body aircraft at lower costs than a wide-body but with fewer window places. Airlines have proved incredibly effective at encouraging people to travel. Even the doubling of passenger air journeys in the first ten years after the introduction of jets, to around 200 million per year in 1967, seems like small fry today. Despite occasional blips, the trend has continued inexorably upward, with annual passenger numbers, at the time of writing, exceeding four billion per year. Would ocean liners and trains have been able to cope with all that traffic?

ABOVE: *Polish-based Jan Kallwejt Studio has produced many high-quality and innovative designs, but the 2011 map for Spanair is a thing of beauty. Its aesthetic balance and simple vignettes hark back to the glory days of airline map production.*

OPPOSITE TOP AND BOTTOM RIGHT: *Exponents of diagrams since 1968 (p.110), Air Canada's current service maps take this to its ultimate conclusion in 2017.*
BOTTOM LEFT: *An Air Mauritius diagram of 2018 contrasts with a hub-and-spoke style.*

Maps continue to feature in airline timetables and publicity, but, sadly, the connoisseur will have to search hard for examples that match the style and panache of previous decades. There are still vibrant, creative designs, especially from less-established airlines with territory to grab. Brash maps shout their message (pp. 136–137), colorfully embellished ones capture attention (pp.138–139), and evocative ones show that the world is a smaller place than it used to be (p. 143). There are diagrams with clever twists (pp. 140, 142), and intricately complex ones that must have taken many hours to create (p. 141). The airline map seems here to stay and, who knows, perhaps another golden age of mapping might be just over the horizon.

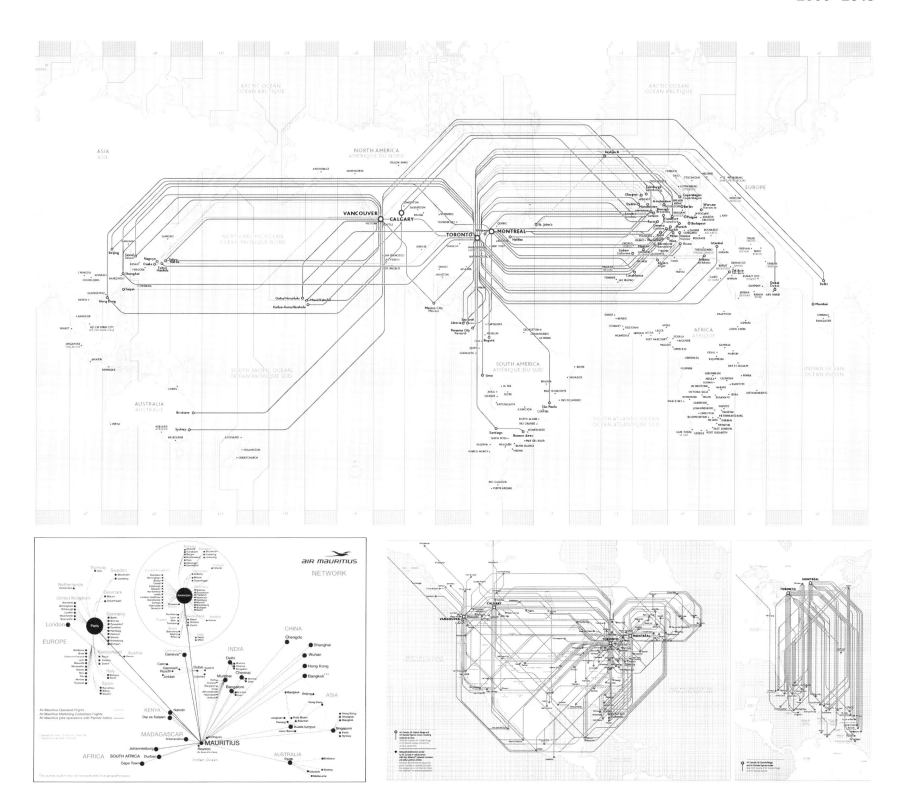

Where shall we fly you to?

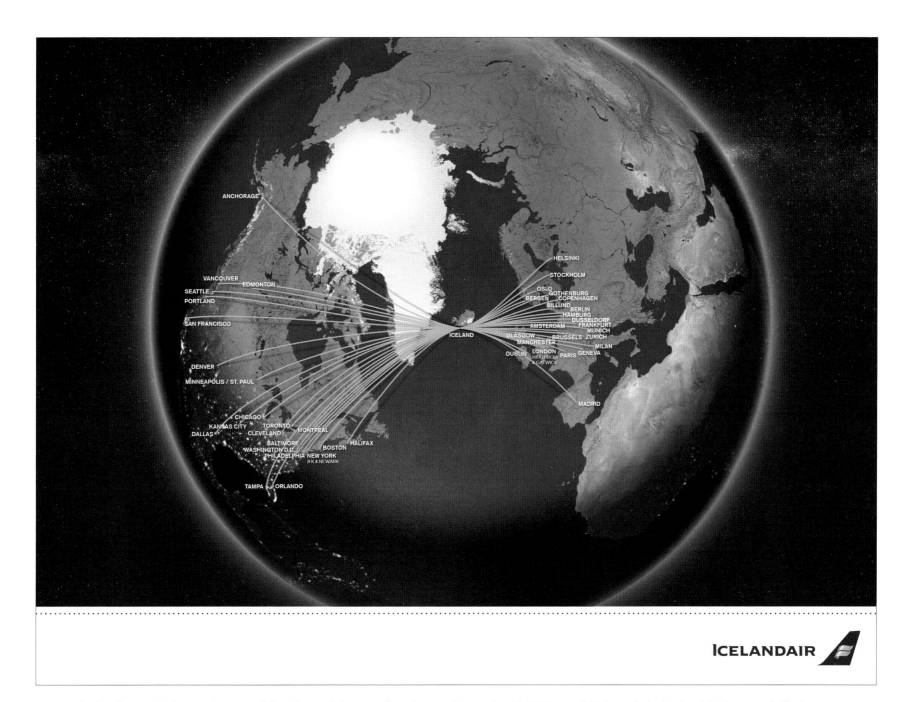

OPPOSITE: *Proving that even in the age of apps, well-thought-out design can still produce excellent results, this 2018 map of Air Europa's destinations is both neat and effective.*
ABOVE: *Icelandair is not the first to center the Earth on its country, but this final image in the collection, from 2018, for one of the world's smallest countries, is so gracefully executed that it takes the prize for sheer chutzpah and élan. The neatly radiating routes give the airline center stage.*

Index

Bibliography

Websites: Aviation History

Airmail History in Pictures: about.usps.com/who-we-are/postal-history/airmail-history-in-pictures.pdf
The World's Airlines History: Past, Present & Future: airlinehistory.co.uk
America by Air: airandspace.si.edu/exhibitions/america-by-air
The History of Flight: century-of-flight.net
The History of Flight: centennialofflight.net/hof/index.htm

Websites: Maps and Timetables

Airline Maps: airlinemaps.tumblr.com
Airline Memorabilia: airline-memorabilia.blogspot.com
Airline Timetable Images: timetableimages.com
CityLab (article): citylab.com/design/2014/08/why-airline-maps-all-look-the-same/375692
Transit Maps: transitmap.net

Books and Print Articles

Bell, Dana, Smithsonian Atlas of World Aviation, HarperCollins, New York, 2008.
Budd, L. C., "On being aeromobile: airline passengers and the affective experiences of flight," Journal of Transport Geography 19: 1010–1016, 2011.
Eskilon, S. J., Graphic Design: A New History, Lawrence King, London, 2007.
Gerstein, Joanne, Fly Now, National Geographic, Washington, DC, 2007.
Hühne, M. C., Airline Visual Identity, 1945–1975, Callisto, Berlin, 2015.
Jarvis, Paul, Mapping the Airways, Amberley Publishing, Stroud, Gloucestershire, England, 2016.
Larsson, Bjorn, Time Flies: Timetables from All Over the World, Trafik-Nostalgiska Förlaget, Stockholm, 2011.
Hadaway, Nina, The Golden Age of Air Travel, Shire Publications, Oxford, 2013.
Harper, Tom, Maps and the 20th Century: Drawing the Line, British Library, 2016.
Monmonier, Mark, The History of Cartography, Volume 6, University of Chicago Press, 2015.
Ovenden, Mark, Transit Maps of the World, Penguin Books, New York, 2015.
Roberts, M. J., Underground Maps Unravelled: Explorations in Information Design, Wivenhoe, UK, 2012.